P9-ECN-334

Gumhalter

Power Supply Systems in Communications Engineering

Power Supply Systems in Communications Engineering

Part I Principles

By Hans Gumhalter

Siemens Aktiengesellschaft
John Wiley and Sons

Title of German original edition:
Stromversorgungssysteme der Kommunikationstechnik
Hans Gumhalter
Siemens Aktiengesellschaft 1983
ISBN 3-8009-1374-7

Deutsche Bibliothek Cataloguing in Publication Data

Gumhalter, Hans:
Power supply systems in communications engineering/Hans Gumhalter. – Berlin; München: Siemens Aktiengesellschaft; Chichester: Wiley

Dt. Ausg. u.d.T.: Gumhalter, Hans: Stromversorgungssysteme der Kommunikationstechnik

Pt. 1. Principles. – 1984.
ISBN 3-8009-1379-8 (Siemens Aktienges.)
ISBN 0 471 90290 X

Library of Congress Cataloging in Publication Data:

Gumhalter, Hans.
Power supply systems in communications engineering. Includes index.
Contents: pt. 1. Principles.
1. Telecommunication systems—Power supply. I. Title.
TK5102.5.G85 1984 621.38 83-13666

ISBN 0 471 90290 X

British Library Cataloguing in Publication Data:

Gumhalter, Hans
Power supply systems in communications engineering.
Pt. 1: Principles
1. Telecommunications 2. Electric apparatus and appliances—Power supply
I. Title II. Stromversorgungssysteme der Kommunikationstechnik. *English*
621.38 TK5102.5

ISBN 0 471 90290 X

Filmset by Pintail Studios, Ringwood, Hampshire
Printed in Great Britain by the Pitman Press Ltd., Bath, Avon.

Foreword

An important part of any communications system is its power supply system. The smooth running of all communications depends directly on the quality of the power supply and thus on the operational reliability of the ever more complex equipment and devices used for the purpose.

It is the intention of this book to explain present-day technology in the supply of power for telecommunications comprehensively and in detail. As this field now finds semiconductor components indispensable, particular attention is devoted to them in a decidedly practical description. This also applies to the subject of automatic control. The reader can thus normally dispense with specialist literature on these subjects.

The individual chapters are constructed and written so that the book will be useful for those readers seriously wishing to become involved in the subject as well as for those having to deal directly with the equipment such as in planning, installation, commissioning and servicing.

A detailed discussion of telecommunications power supply equipment is being reserved for a section of its own to appear later; it will review the state of affairs with systems developed within the Company and introduced with the user using thyristor and transistor-controlled converters and also explain the circuit correlations in a largely general manner. The second part also deals with the subjects of use of batteries, earthing and protective measures.

The selection of material and method of presentation are based on knowledge and experience gained in the context of training measures for power supply systems.

Munich, February, 1984

Siemens Aktiengesellschaft

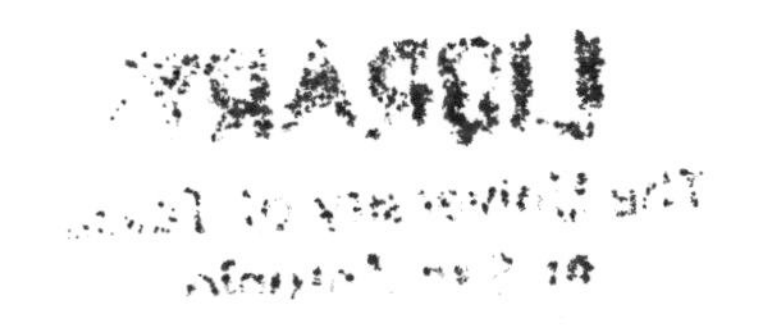

Contents

1 Outline and Historic Development

The power supply of any telecommunication system forms the link between the national power supply (mains) and the communications system itself. Thus not only must it be matched to the parameters of the mains but also fulfil the requirements of the communications system and its energy stores, i.e. batteries. This is achieved by transforming the voltages of the mains, or standby power supply system, when requirements such as magnitude tolerance, purity, etc., have to be taken into consideration.

Measures need to be taken to maintain the operational reliability of the communications system in the event of temporary loss of power or complete mains failure.

Telecommunication power supply systems are classified according to:

▷ type of power supply system,
▷ type of automatic control for the individual devices, units and equipment.

Important aspects of a telecommunications supply include:

▷ observance of the tolerance range for the supply voltage, namely:
  between no load and rated load,
  with surges in load,
  with mains voltage fluctuations,
  with changes in mains frequency;
▷ maintenance of the purity of the direct voltage in accordance with the specifications, i.e. the limit values for superimposed alternative voltage must not be exceeded.
▷ a power supply free as far as possible from interruption;
▷ presence of sufficient monitoring, protective, limiting and signalling devices;
▷ simplicity of extension;
▷ economy;
▷ small dimensions and light weight;
▷ ease of installation and maintenance;
▷ design to relevant specifications and regulations.

Four basic types of power converter are used for telecommunications power supply systems in the control or conversion of electrical power with the aid of power electronics components (Fig. 1.1).

① With *rectifiers* an alternating current can be converted into a direct current. Energy flows from the alternating current system to the direct current one.

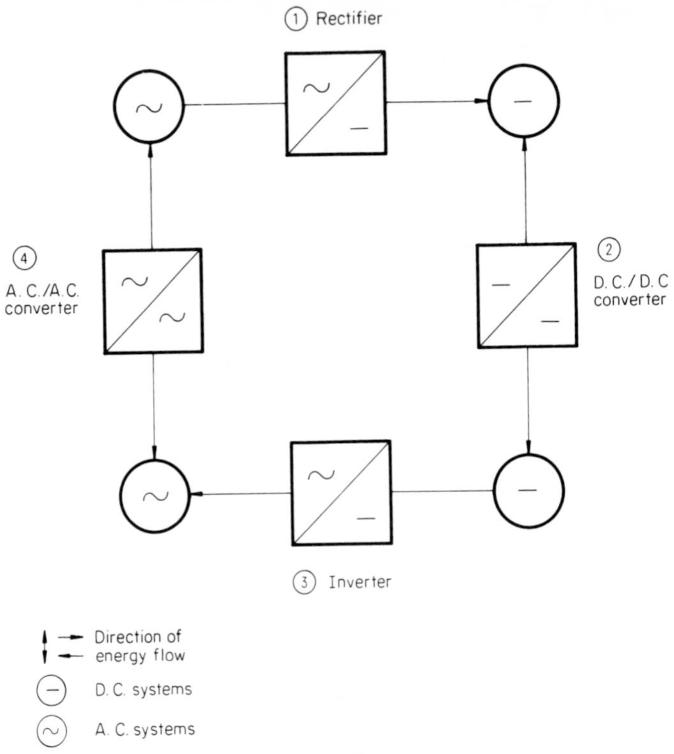

Fig. 1.1 Functions of the basic types of power converter

② *D.C./D.C. converters* enable direct current of a given voltage and polarity to be converted into direct current of another constant or changeable voltage and/or polarity. In today's power supply systems d.c./d.c. converters are often used to obtain component voltages from an 'energy transport voltage' of, for example, 48 V.

③ *Inverters* (d.c./a.c. converters) are used for converting direct current into alternating current. Energy flows from the direct current system to the alternating current one.

④ *A.C./A.C. converters* enable alternating current of a given voltage, frequency and number of phases to be converted into alternating current of another voltage and/or frequency and/or number of phases.

The typical elements of a power converter with the energy conversion flow are shown in Fig. 1.2.

Figure 1.3 shows the outline of typical telecommunications power supply systems, while Fig. 1.4 shows an overall view of such a system.

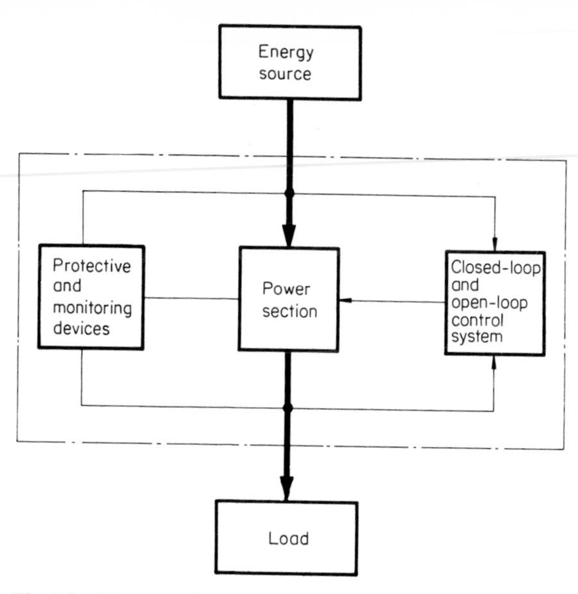

Fig. 1.2 Elements of a power converter

① mains supply and standby power supply system (emergency unit with internal combustion engine);
② mains distribution switchboards or mains switch panels;
③ a central power supply;
④ a decentralized power supply.

① Mains supply and standby power supply systems

The necessary electrical power is usually taken from the national mains. As there exists the possibility of failure with this mains supply a standby fixed or mobile power supply is used where necessary. Both the mains supply and the standby power supply system provide a single-phase or three-phase alternating voltage.

Trouble other than power failure can occur with the mains supply, e.g.:

▷ failure of individual phases,
▷ excessive voltage or frequency fluctuations,
▷ excessive harmonics.

Further arrangements, apart from standby power supply systems, which bridge disturbances in the mains supply are to be found in ③ 'Central power supply'.

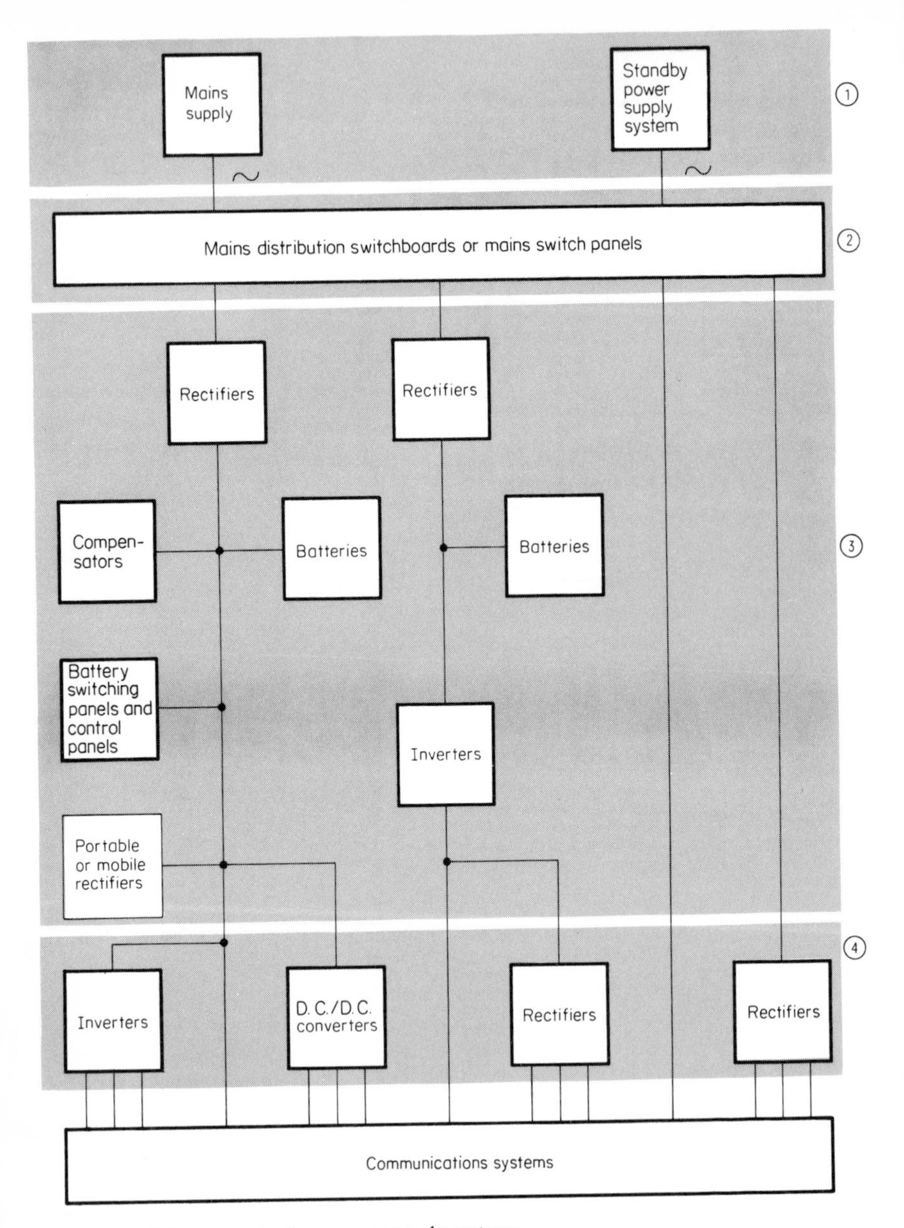

Fig. 1.3 Telecommunications power supply systems

Fig. 1.4 Telecommunications power supply system

② **Mains distribution switchboards or mains switch panels**

Mains distribution switchboards and mains switch panels are used for the distribution, switching and measurement of the mains voltage and for providing fuse protection. While power supply equipment of lower output can be linked with the mains supply via a mains distribution switchboard, rectifiers of greater output power must be connected to a mains switch panel.

③ **Central power supply**

A distinction is made between a central and a decentralized power supply of a communications system. A central power supply will be considered first.

Rectifiers

The rectifiers in a central power supply system feed the communications system, inverters, batteries and d.c./d.c. converters. These rectifiers are with thyristor power section (thyristor-controlled rectifiers) and convert the alternating current of the mains supply or standby power supply systems into direct current. The rectifiers together with batteries and possibly any standby power supply systems must be capable of providing an uninterruptible supply of direct current for the communications system.

The most important modes of operation in the case of rectifiers are:

▷ Rectifier mode (without battery)
  In the event of a power failure the communications system does not receive any supply voltage.
▷ Parallel mode (with battery)
  Rectifiers, battery and communications system are connected in parallel. In this mode a distinction is made between standby parallel mode and floating mode.
▷ Changeover mode (with battery)
  In the event of power failure the supply is switched from mains to battery operation without interruption.

The essential components of thyristor-controlled rectifiers are:

▷ mains transformer,
▷ thyristor set,
▷ filter,
▷ closed-loop and open-loop control,
▷ protective and monitoring devices,
▷ signalling system.

The normal ratings for the d.c. output voltage provided by the rectifiers are 48 and 60 V.

Thyristors, as final control element, together with their associated control equipment, ensure that changes in the power consumption of the communications system or fluctuations in the mains voltage or frequency result in the d.c. output voltage from the rectifiers varying only within a permissible tolerance.

Monitoring devices on the direct voltage side protect the equipment against excessive over- or undervoltages which would interfere with its proper working.

There are electronic mains monitoring devices to protect rectifiers operating on three-phase mains. These monitoring devices detect the failure of one or more phases, any mains under- and overvoltage, as well as any asymmetry, and switch off the relevant rectifier. The device automatically switches the rectifier back into circuit when normal conditions have been restored.

Thyristor-controlled single-phase rectifiers of up to about 25 A rated current have a two-pulse converter circuit and three-phase rectifiers from 25 to 1000 A have a six-pulse one. In rectifiers with a rated current from 500 to 1000 A (GR10) the unwanted mains reaction is still further reduced, compared with the six-pulse arrangement, by switching to a twelve-pulse one (paired operation of rectifiers).

All thyristor-controlled rectifiers are fitted with noise suppression as standard to 'VDE 0875/DIN 57875' – at least to degree of radio interference *N*.

Batteries

If the rectifier fails to provide a supply, a continued supply to the communications system can be ensured with the aid of batteries (accumulators). Sometimes a standby power supply system, mentioned in ①, can also be used.

Portable and mobile rectifiers

Portable and mobile power supply equipment with the technical characteristics of stationary rectifiers can replace the fixed system.

Inverters

Inverters (d.c./a.c. converters) are used when a communications system requires an alternating current supply. In this way the demand on the central power supply for an alternating voltage free from interruptions can be met using batteries and rectifiers.

Compensators

In the event of a power failure a compensator, fed by the battery, supplies a gradually rising booster voltage in series with the battery voltage. Thus, despite a falling battery voltage, the supply voltage for the communication system remains within the permissible tolerance, even with a power failure.

Battery switching panels and control panels

A distinction is made between *battery switching panels with control* and *battery switching panels.*

Battery switching panels with control contain not only the power circuits but also the whole control, monitoring and signalling systems for the power supply. A maximum of two batteries can be connected to them. No *central control panel* is necessary with these power supply systems.

Battery switching panels contain only the power circuits. Each battery must have its own battery switching panel. A central control panel is also necessary for each system. Arranged within this panel are the complete control, monitoring and signalling systems for the power supply.

Small to medium-sized power supply systems can also be constructed without battery switching panels, control panels and compensators. Some systems are even supplied without a battery. The ultimate design depends on what communications system the power supply is being planned for, what type series of rectifiers has been decided upon and what reliability requirements are to be observed.

④ Decentralized power supply

Normally a number of low supply voltages (component voltages of, for example, 5, 12 and 24 V) with narrow tolerances and differing polarities are required for modern communications systems. These are obtained from the central supply voltage (in this case energy transport voltage) of, for example, 48 V.

It would be uneconomic to provide the various component supply voltages centrally. With small supply voltages the current would be larger. The voltage drop on the lines would therefore also increase, leading to a current distribution system with disproportionately large line cross-sections. It would also be necessary for each component supply voltage to have its own supply line.

A decentralized power supply system enables the power supply to also expand when a communications system is extended. There is also the factor that the effect of disturbance is more confined.

The output voltage is controlled by power transistors in switching mode at a clock frequency of 20 kHz or higher or, less often, as longitudinal controllers.

When the power supply is decentralized different systems exist for providing component voltages:

Systems for input d.c. voltage

▷ Provision of direct voltages with d.c./d.c. converter,
▷ Provision of alternating voltages with inverter.

Systems for input a.c. voltage

▷ Provision of direct voltages with rectifier.

Power supplies units with input d.c. voltage

▷ D.C./D.C. converter

Direct current converters are also called:
d.c. chopper controllers or regulators,
d.c. voltage transformers,

A distinction is made between three types of converters:

▷ single-ended blocking converters,
▷ single-ended flow converters,
▷ push–pull flow converters.

From the rectifier of the central power supply the direct voltage arrives at the d.c./d.c. converter of the decentralized power supply system. This voltage is chopped into a square-wave voltage by a rapid-switching transistor. The square-wave voltage passes through a transformer and is then rectified and filtered. The d.c. output voltage is stabilized by a regulating circuit, which modifies the duty cycle of the transistor switch (pulse-width control).

The tolerance limits for supply voltages to the d.c./d.c. converter are very wide. Direct current converters process input d.c. voltages from 40 to 75 V. The d.c. output voltages, as already explained, are held constant at the desired component voltage levels with a narrow range of tolerance.

▷ Inverter
In certain cases decentralized power supply facilities also include inverters which provide alternating voltage from the energy transport direct voltage from the central power supply. This alternating voltage is needed to feed peripheral devices such as page printers and memories.

Power supplies units with input a.c. voltage

▷ Rectifiers
Rectifiers with transistor power section (transistor-controlled rectifiers) are linked either with the inverters of the central power supply or directly with the mains supply. Here they are 'cycled' elements of the switching mode power supply or devices with a longitudinal regulator.

There is alternating voltage at the input of the switching mode power supplies which is rectified with a mains input rectifier circuit. There then follows a d.c./d.c. converter circuit.

Transistor-controlled rectifiers with a switching frequency of more than 10 kHz are fitted with noise suppression to 'VDE 0871/DIN 57871' – limit value class B.

Power supply systems independent of the mains

Power supply systems for smaller outputs independent of the mains occupy a special position. They are frequently housed in shelters.

The following sources of primary current are used:

units with internal combustion engine,
small steam turbines,
thermal generators,
solar generators and
wind-driven generators.

Fuel consumption and maintenance costs can be reduced by means of hybrid systems. Wind-driven generators and/or solar generators are combined, for example, with diesel generating sets and batteries (Figs 1.5 and 1.6).

Historic development

This introduction to the supply of power for telecommunications systems will be rounded off with a review of the prominent steps which led to its development. It can, moreover, through the inserted comments, facilitate the understanding of new techniques and technologies.

Galvanic or *voltaic cells* were the first sources of electrical energy.

The development of a power supply for a telecommunications system began with the invention of the telephone in 1861 by *Philipp Reis* and the improvements made by *Alexander G. Bell* and *David E. Hughes* (1876).

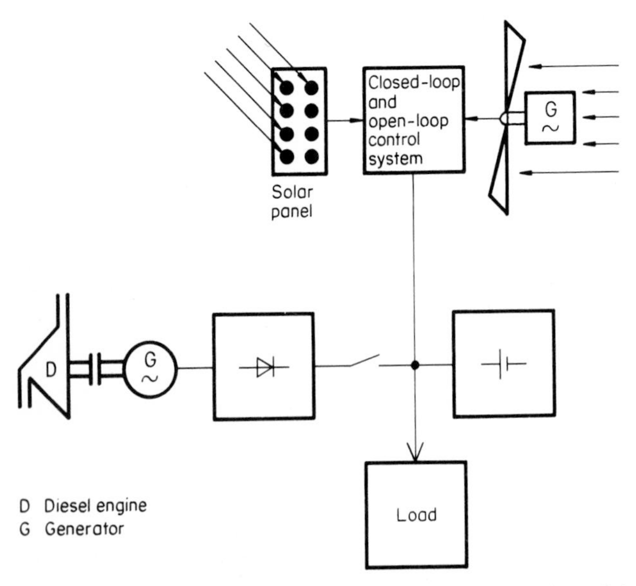

Fig. 1.5 Block diagram of a power supply system independent of the mains supply with diesel generating set, solar generator and wind-driven generator

Fig. 1.6
Power supply system independent of the mains with diesel generating set, solar generator and wind-driven generator

The first *public telephone network* in Germany with a hand-operated exchange was put into operation in Berlin as early as 1881. This was a *local battery system* (LB system) in which each subscriber extension had its own power supply in the form of a primary cell (*dry battery*).

For reasons of economy and operating reliability a switch was made from the LB system to the *central* or *common battery system* (CB system) on introduction of the first automatic systems (around 1900). With this system the current sources were located centrally with the switching system and, instead of primary cells, secondary cells (lead accumulators or storage batteries) came into use (invented in 1859 by the Frenchman *G. Planté* and based on the work of *Sinsteden* who had introduced lead into secondary batteries in 1854).

Around the same time as the introduction of automatic systems a decision was made for today's public *alternating current supply*.

The mode of operation with the CB system was initially with two batteries, also called the charge–discharge mode (battery mode, Fig. 1.7). While battery 2 supplied the system (communications system) battery 1 was being simultaneously charged by a rotating converter (a.c. motor coupled with d.c. generator). At certain intervals

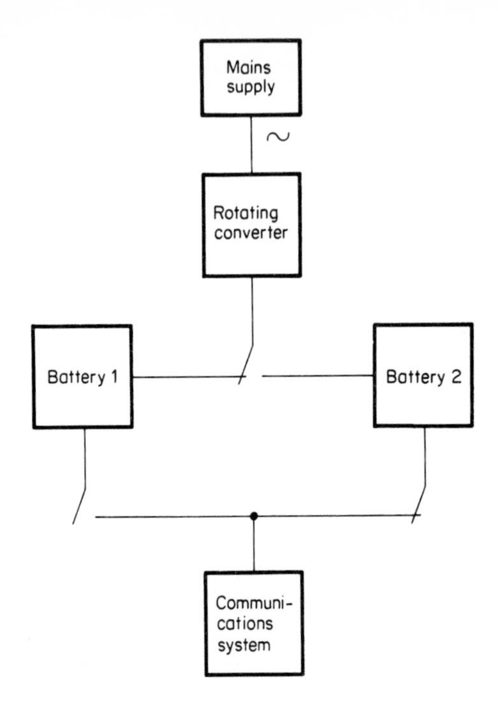

Fig. 1.7
Telecommunications power supply – two-battery mode with rotating converter

of time there was a switchover so that battery 1 then fed the communications system and battery 2 was recharged by the converter.

Later *parallel mode* (floating mode) replaced operation with two batteries. For reliability two batteries are normally connected in parallel. At first a *rotating converter* still fed the battery and communications system. The converter, battery and communications system were always connected in parallel (Fig. 1.8 ①).

Invention of the *mercury arc rectifier* (1902) by *P. Cooper-Hewitt* replaced the rotating converter, which was well known to be inefficient and required a lot of maintenance. As with today's thyristor-controlled rectifiers it was already possible with the mercury arc rectifier to keep the d.c. output voltage constant by modifying the firing point (Fig. 1.8 ②).

Arrival of the polycrystalline semiconductor around 1930 permitted construction of the first 'dry rectifier' (*copper oxide rectifier*) for the power range up to 60 V/3 A (Fig. 1.8 ③). The mercury arc rectifier, still to be encountered today, is used for higher powers.

Around 1934 the *selenium rectifier* (selenium diode) was developed – like the copper oxide rectifier also based on polycrystalline semiconductor material. The selenium rectifier with its higher specific load capacity replaced the copper oxide rectifier. It is in fact more sturdy than the mercury arc rectifier, but is not suitable

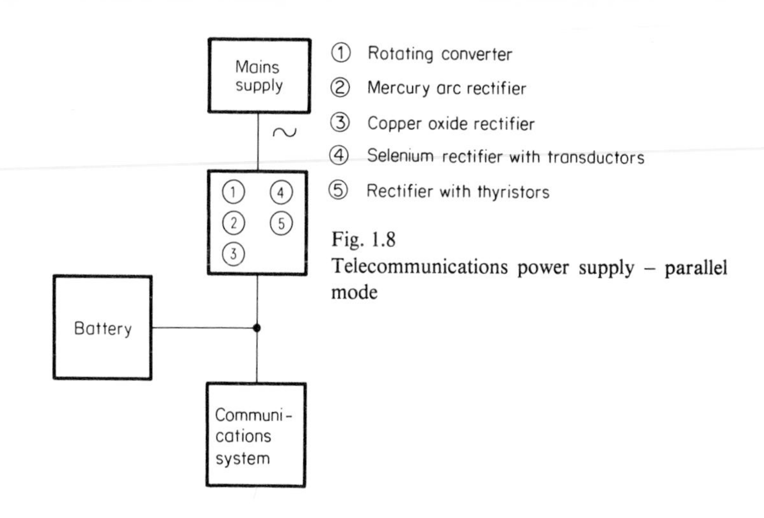

Fig. 1.8
Telecommunications power supply – parallel mode

for maintaining a constant output voltage. For this reason the mercury arc rectifier was preferred to the selenium rectifier until around 1949 when there were suitable control elements for selenium rectifiers – *transductors*, also called *magnetic amplifiers* (Fig. 1.8 ④). There thus came into being the first *magnetically controlled* rectifiers.

The importance of the mercury arc rectifier declined rapidly from that time on.

The selenium rectifier is still used today for certain applications in telecommunications power supply systems.

The magnetically controlled rectifiers already had an *automatic control circuit* with magnetic regulating inductors for control. The automatic control circuit contains units working solely on a magnetic basis, e.g. magnetic controllers, transductors, etc. (Fig. 1.9 ①).

With mastery of the behaviour of single crystals there appeared in 1948 the first single-crystal semiconductor devices – *germanium diode* and *germanium transistor* (invented by *J. Bardeen, W. H. Brattain* and *W. Shockley*, USA). The technology now existed for the development of *transistor controllers* instead of magnetic controllers in magnetically controlled rectifiers.

Magnetically controlled rectifiers with transistor controllers appeared around 1960 (Fig. 1.9 ②), providing better control behaviour. Such rectifiers are still being built today without any major modifications.

Figure 1.10 shows an example of a magnetically controlled rectifier in *changeover mode* with battery tap. In normal operation a main rectifier supplies the communications system, while the battery is fed by the main and additional rectifiers working together. Only in the event of a power failure is the battery connected to

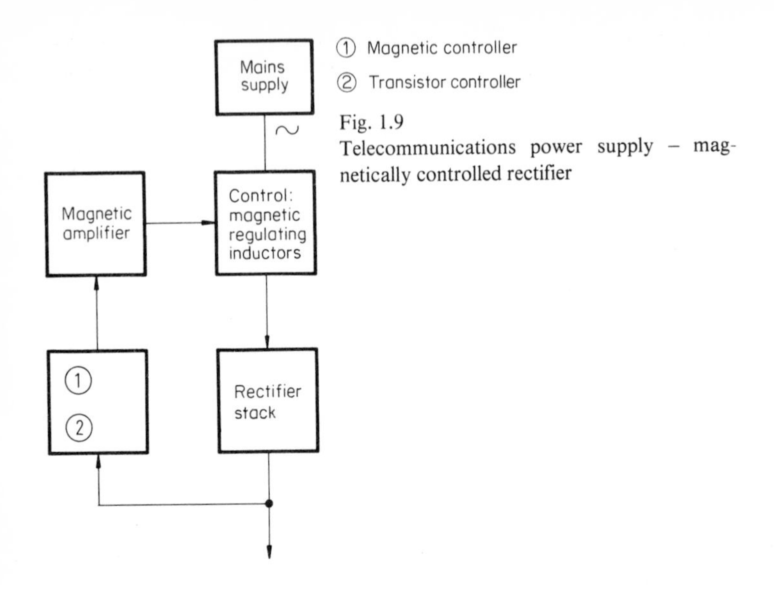

Fig. 1.9
Telecommunications power supply – magnetically controlled rectifier

the communications system with the aid of the battery discharge contactor. During the switching time of the contactor which is some 100 ms in duration, 26 cells of the battery provide the supply to the communications system via the tapping diode (tapping valve), thereby guaranteeing an uninterrupted transition from mains to battery operation.

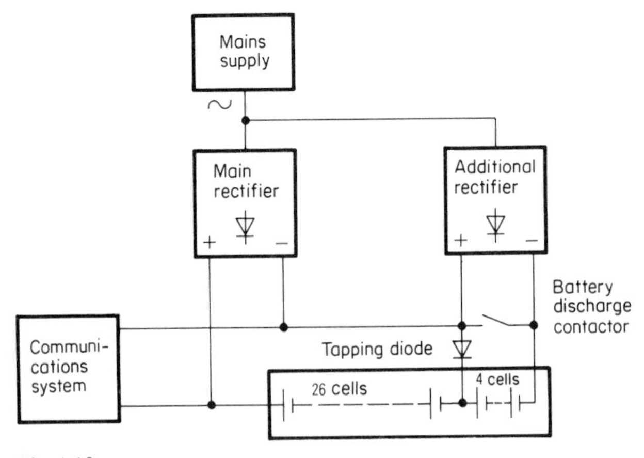

Fig. 1.10
Telecommunications power supply – changeover mode with battery tap for 60 V communications systems

For smaller and medium-sized Private Automatic Branch Exchange (PABX) systems there were also *uncontrolled* and *phase-controlled* rectifiers in addition to the magnetically controlled ones.

It is well known with uncontrolled rectifiers that the d.c. output voltage drops as the load increases and is also dependent on frequency and mains voltage fluctuations (Fig. 1.11).

Uncontrolled rectifiers are suitable for current strengths of about 0.7 to 3 A.

Phase-controlled rectifiers were 'self-controlling'. Like magnetically controlled rectifiers they kept the d.c. output voltage constant despite mains voltage fluctuations and changing load conditions throughout the communications system. Changes in frequency, however, do influence the d.c. output voltage (Fig. 1.12).

Phase-controlled rectifiers were used for current strengths from 1.5 to 25 A.

From around 1955 there have existed single-crystal semiconductor devices based on silicon, namely the *silicon diode* and the *silicon transistor.* The silicon diode has a higher reverse voltage, greater load capacity and a steeper characteristic than the selenium diode. That is why with magnetically controlled rectifiers the rectifier stack was later designed with silicon diodes.

In 1956 in the United States, *Moll, Tanenbaum et al.* developed a variant of the transistor, which can be considered a *silicon-controlled rectifier*. In West Germany, by analogy, the hot-cathode gas-filled tube (thyratron) was at first called the silicon-hot-cathode gas-filled tube, but later since the early 1960s it has been called a *thyristor.* It has brought about a turning point in rectifier technique and is today the predominant component in power electronics.

Figure 1.8 ⑤ shows the principle of a thyristor-controlled rectifier in parallel mode.

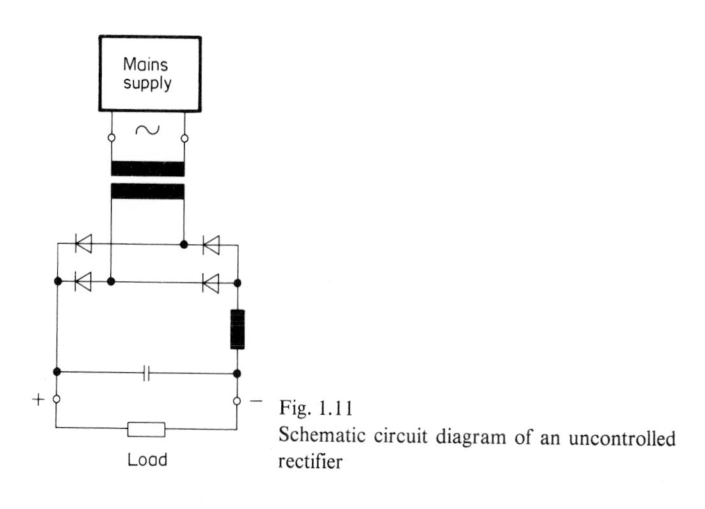

Fig. 1.11
Schematic circuit diagram of an uncontrolled rectifier

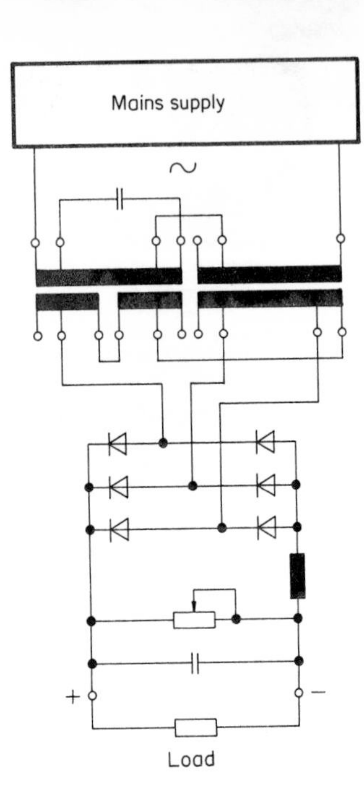

Fig. 1.12
Schematic circuit diagram of a phase-controlled rectifier

When in 1965 medium-sized PABX systems (*telephone system 400 E*) came out using ESK crosspoint switching technique, thyristor-controlled rectifiers could then already be used for the power supply. They are still available today with these systems for current strengths of 3, 5, 10 and 16 A.

In 1973 a 48 V/40 A *thyristor-controlled rectifier* came onto the market for the KS 3000 E communications system (large-sized PABX).

1974 already saw thyristor-controlled rectifiers for up to 1000 A being built.

Since 1976 *integrated circuits* have been available for rectifiers.

The first *switching-mode power supplies* came out in 1977 after power transistors were developed. These are now also being used in new communications systems.

2 The Requirements Communications Systems Place upon Telecommunications Power Supply Systems

Table 2.1 contains data concerning the characteristics of power supply systems relevant to communications systems.

Figure 2.1 shows a view of a modern communications system (Digital Electronic Switching System EWSD).

2.1 Magnitude of the Direct Voltages

Table 2.1, line 1, shows the *rated voltages* for communications systems. A distinction is made between 24, 48 and 60 V systems. Modern systems are usually 48 or 60 V, the *positive pole* normally being *earthed.*

The choice of voltages also depends on the safety regulations of the country in question. In West Germany direct voltages above 120 V and alternating voltages above 50 V are classed as 'dangerous contact voltages'.

On the output of any telecommunications power supply system is the *central supply voltage* for the communications system (load) and for the battery (Fig. 2.2, cf. also Table 2.1, line 2).

Examples of *battery voltages* are:

2.23 V/cell (trickle charging voltage),
2.33 V/cell (charging voltage),
2.7 V/cell (initial charging voltage).

The central supply voltage and the *operating voltage* differ in the voltage drop ΔU on the distribution lines.

The voltages for conventional communications systems are normally fed direct. Modern systems, on the other hand, additionally require *decentralized supply voltages*, also termed component or electronic voltages (e.g. 5, 12, 24 V). These secondary voltages are obtained from the central supply voltage via a d.c./d.c. converter. The central supply voltage is then called the energy transport voltage.

Table 2.1
Data concerning the characteristics of power supply systems relevant to communications syst

	Communications systems		
	EMD	EMD[2])	EMD[3])
1 Rated voltage in V	48	60	60
2 Central supply voltage in 'normal' mode in V	51	62	62
3 Continuously permissible tolerance for operating voltage measured at the system's subassemblies or components in V	45 to 54	58 to 64	58 to 6
4 Top critical voltage measured at the system's subassemblies or components ($t = 0$ s) in V	$\leqq 75$	$\leqq 90$	$\leqq 90$
5 Frequency-weighted interference voltage in accordance with A-filter curve in mV (CCITT)	$\leqq 2$	$\leqq 2$	$\leqq 2$
6 Superimposed alternating voltage (without frequency weighting) in mV	$\leqq 600$	$\leqq 600$	$\leqq 600$
7 Degree of radio interference	Degree of radio interference N to VDE 0875/DIN 57875[1])		
8 Maximum voltage drop ΔU overall in V (design basis)	2.4	Without compensator 2.4 With compensator 3.4	2.4

EMD Noble-Metal Uniselector Motor Switch-Switching System
ESK Crosspoint Switching System
EWSA Analogue Electronic Switching System
EWSD Digital Electronic Switching System
EMS Electronic Modular System (PABX)
KS Communications System (PABX)

2.2 Tolerance for Direct Voltages

For each communications system the *tolerance* is shown for the operating voltage given in Section 2.1 (Table 2.1, line 3). The *bottom, continuously permissible voltage* is determined by the necessary reliability against a wrong connection when putting through the call and by maintenance of the connected call. The *top, continuously*

ESK 10 000 E	KS 3000 E	EWSA	EWSA	EWSD	EMS
48	48	48	60	48	48
51	51	Direct load 51	Direct load 62	56	53.5
		D.C./D.C. converter 58	D.C./D.C. converter 67		
44 to 53	44 to 54	Direct load 44 to 53	Direct load 57 to 64	Direct load 44 to 58	Direct load 42 to 58
		D.C./D.C. converter 40 to 75			
$\leqq 60$	$\leqq 56$	$\leqq 60$	$\leqq 75$	$\leqq 80$	$\leqq 70$
$\leqq 2$	$\leqq 0.5$	Direct load $\leqq 2$			$\leqq 0.5$
		D.C./D.C. converter – no conditions			
$\leqq 600$	$\leqq 150$	$\leqq 195$			$\leqq 150$
2.6	2.6	2.7	Without compensator 2.4	1.8[4])	2
			With compensator 3.4		

[1]) If power supply units which cause continuous interference of more than 10 kHz are used, stricter conditions of limiting value class B to VDE 0871/DIN 57871 will apply to these units.
[2]) Predominantly for systems of the Federal German Postal Administration (DBP) with 30-cell battery.
[3]) Predominantly for systems in non-European countries with 31-cell battery.
[4]) Respectively 2.8.

permissible voltage depends on the heat generated by the components (relays, selectors, etc.).

The *top critical voltage* must not be exceeded, even for a short time, as components may be destroyed (e.g. semiconductor elements, Table 2.1, line 4).

In the case of modern communications systems the observance of a d.c. output

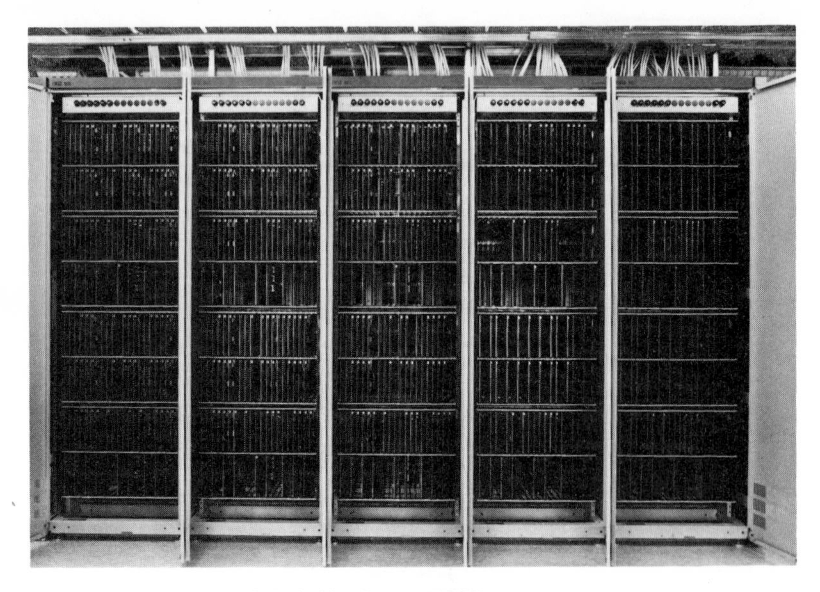

Fig. 2.1 Digital Electronic Switching System EWSD

voltage tolerance of $\leqq \pm$ 0.5%, possible nowadays with rectifiers, is necessary only with respect to a battery in parallel.

A tolerance range of $\pm$ 4% must be allowed in the case of sudden changes in load or mains supply voltage because of the dynamic behaviour of rectifiers.

With the operating mode described in Chapter 3, the standby parallel mode with reducing diodes, the current-dependent voltage drop results in a tolerance of $\pm$ 2% for the operating voltage instead of $\pm$ 0.5%.

When the *battery is discharging* the operating voltage depends on the battery's characteristic curve. If a voltage with a narrower tolerance is required (e.g. 61 V $\pm$ 1%), a compensator must be added.[1])

Direct voltages are monitored by voltage-monitoring systems built either into the rectifiers or into the control devices of the battery switching panels or control panels.

An alarm is triggered if there is a drop below the set values for the bottom voltage limits. An alarm is also triggered and the rectifier switched off if the top voltage limit is exceeded, e.g. due to a fault in the control system.

---

[1]) Compensator only for 60 V systems of the Federal German Postal Administration.

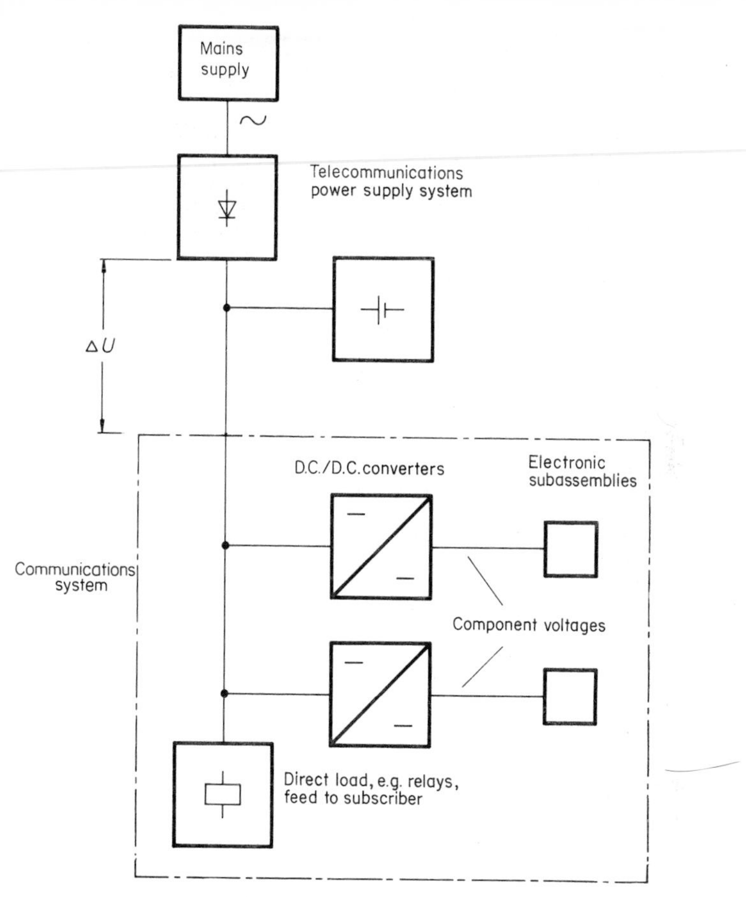

Fig. 2.2 Telecommunications power supply system with battery and load

2.3 Purity of Direct Voltages

The most important definitions for the purity of the direct voltage are given in DIN 41750, Sheet 4, Supplement, and DIN 41755, Sheet 1.

2.3.1 Superimposed alternating voltages

The direct voltage taken at the output of a power supply system always has a superimposed alternating voltage which gets into the speaking and hearing circuits of the communications system in various ways. Hum interference is produced depending on the frequency of interfering alternating voltage and on the transmission characteristics of the telephone circuit. Apart from the requirement that the direct voltage must be as constant as possible, its purity is particularly important for the perfect transmission of information.

The superimposed alternating voltage is made up of a composition of frequencies. These individual frequencies cause varying degrees of interference.

The purity of the direct voltage is defined by international provisions of the CCITT[1]) so that a constant quality is ensured in the case of communications across a number of national or international networks.

A weighting curve has been defined for the degree of interference of the individual frequencies. This is the CCITT 'A-filter curve' (Fig. 2.3).

[1]) *C*omité *C*onsultatif *I*nternational *T*élégraphique et *T*éléphonique (International Telegraphs and Telephones Consultative Committee)

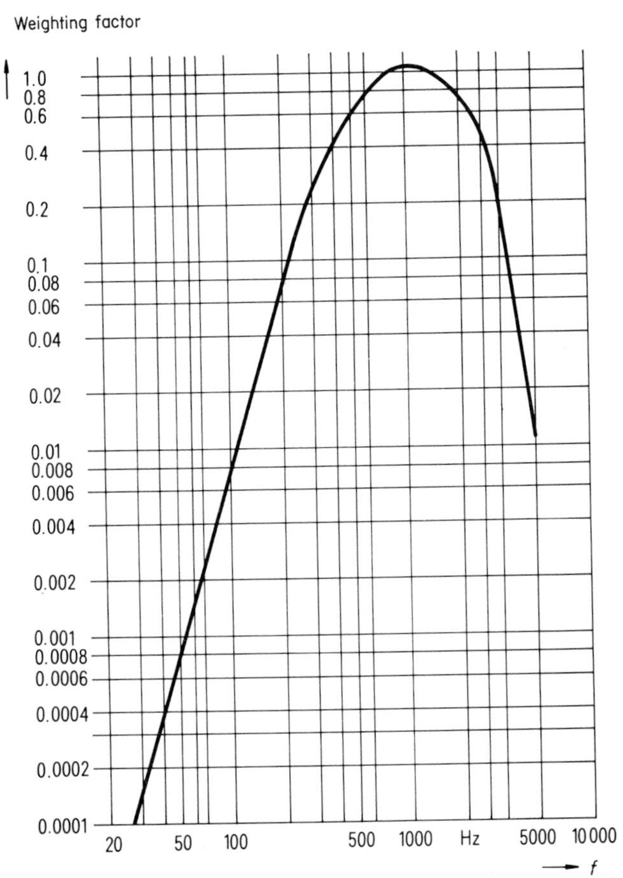

Fig. 2.3 The CCITT A-filter curve

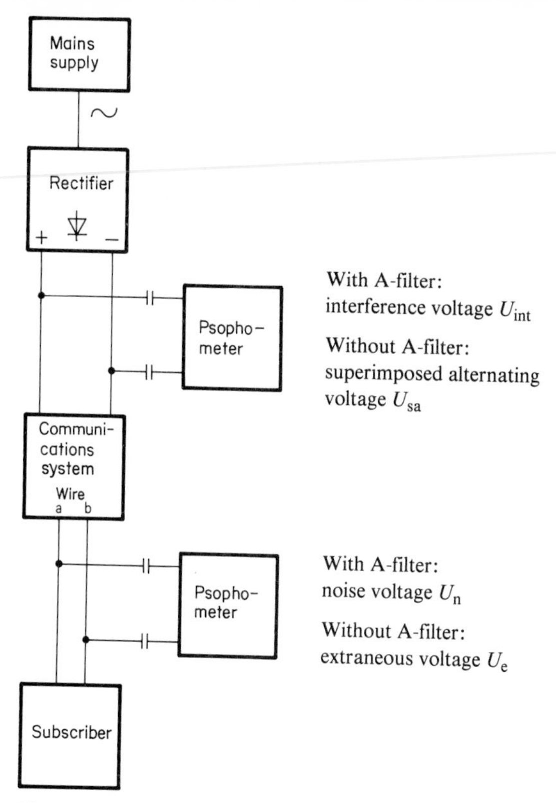

Fig. 2.4
Measuring the superimposed alternating voltage U_{sa}, interference voltage U_{int}, extraneous voltage U_e and noise voltage U_n. The capacitors shown are normally incorporated in modern psophometers

It can be seen that low frequencies produce less interference; if the frequency increases, so does the interference. At frequencies of more than 1 kHz the psophometric weighting factor becomes smaller again and the interference is reduced.

The weighting factor takes into account the effect on the human ear and the electroacoustic characteristics of the telephone receiver. To calculate the interference voltage using the A-filter curve the weighting factor must be multiplied by the superimposed alternating voltage.

A psophometer (circuit noise meter) is used to measure the superimposed alternating voltage (Fig. 2.4). It has two switch positions: 'extraneous voltage' (without the A-filter) and 'A-filter' (noise voltage) positions. In the 'extraneous voltage' position the sum of the superimposed alternating voltages (root mean square r.m.s. values) is indicated. In the 'A-filter' position the weighted alternating voltage is indicated aurally and compensated for by the built-in filter (according to the sensitivity of the ear's threshold of hearing). The uniform weighted A-filter curve is related to a frequency of 800 Hz.

When measuring a distinction is normally made between:

- ▷ the interference voltage U_{int}, the alternating voltage at the power supply output weighted with A-filter,
- ▷ the superimposed alternating voltage U_{sa}, the unweighted alternating voltage at the power supply output,
- ▷ the noise voltage U_n, the alternating voltage weighted with A-filter on the speaking wires of a telephone connection,
- ▷ the extraneous voltage U_e, the unweighted alternating voltage on the speaking wires of a telephone connection.

As an international standard the noise voltage U_n between the speaking wires in any communications system, measured with the circuit noise meter with A-filter, must not exceed 0.2 mV.

Table 2.1 shows the individual permissible maximum values for interference voltage (line 5) and the superimposed alternating voltage (line 6). It follows, therefore, that the maximum permissible value for the interference voltage (or the superimposed alternating voltage) is dependent on the communications system. The different dimensioning of the output filter arrangement in the power supply system takes into account the fact that the inductance of the feeding bridge in a communications system differs. This becomes clear if, for example, a flat-type relay is compared with an ESK noble metal high-speed contact relay. For this reason it is necessary in the case of different communications systems to reduce the maximum permissible interference voltage at the power supply output to 0.5 mV so that the noise voltage within the system does not become greater than 0.2 mV.

The extraneous voltage U_e is proportional to the superimposed alternating voltage U_{sa} as the noise voltage U_n is proportional to the interference voltage U_{int}.

Within the framework of a detailed consideration of the superimposed alternating voltage on the direct current side there now follows an explanation of the terms alternating voltage component, ordinal number, alternating voltage content, etc.

An alternating voltage is superimposed on the direct voltage U_d (arithmetic mean). It consists of sinusoidal components of different frequencies νf,

where

ν is the ordinal number and
f is the frequency.

The ordinal numbers ν of these components are integral multiples of the pulse number p:

$$\nu = kp \qquad \text{with } k = 1, 2, 3, \ldots$$

where

k is the distortion factor.

The ideal alternating voltage component of the ordinal number ν has, at full modulation (control angle $\alpha = 0°$), the r.m.s. value:

$$U_{\nu i} = \frac{\sqrt{2}}{\nu^2 - 1} U_{di}$$

where

U_{di} is the ideal direct voltage.

With partial modulation the values for $U_{\nu i\alpha}$ (ideal alternating voltage component of the ordinal number ν as a function of the control angle α) increase as the control angle increases.

With thyristor-controlled rectifiers it is possible, e.g. in the case of capacitive loading, for the current to have gaps as a function of the degree of modulation. In contrast with the ideally stabilized direct current, the behaviour of current and voltage differs in 'gap mode' (control angle α very large).

The alternating current content w is at a minimum value with full modulation and reaches its maximum value with least modulation. It follows from this that the alternating voltage content w of the direct voltage U_d is primarily dependent on the control angle and increases as the control angle becomes larger.

The ideal alternating voltage content (ideal ripple) $w_{i\alpha}$ is the ratio of the r.m.s. value of the superimposed ideal alternating voltage to the ideal direct voltage:

$$w_{i\alpha} = \frac{U_{si\alpha}}{U_{di}} = \frac{\sqrt{\Sigma U_{\nu i\alpha}^2}}{U_{di}}$$

where

$U_{si\alpha}$ is the superimposed ideal alternating voltage when modulated with the control angle α.

For the ideal alternating voltage content at full modulation:

$$w_i = \frac{U_{si}}{U_{di}} = \frac{\sqrt{\Sigma U_{\nu i}^2}}{U_{di}}$$

the values for pulse numbers 2, 6 and 12, taking into account all even ordinal numbers $\nu = 2$ to 24 with their assigned frequencies, are to be taken from Table 2.2.

It is seen that as the ordinal number ν rises, the quotient $U_{\nu i}/U_{di}$ becomes smaller.

Table 2.2
Ideal superimposed alternating voltage on the direct current side with full modulation (control angle $\alpha = 0°$)

ν	νf	Pulse number p		
		2	6	12
			$\frac{U_{vi}}{U_{di}}$ %	
2	100	47.14	—	—
4	200	9.43	—	—
6	300	4.04	4.04	—
8	400	2.24	—	—
10	500	1.43	—	—
12	600	0.99	0.99	0.99
14	700	0.73	—	—
16	800	0.55	—	—
18	900	0.44	0.44	—
20	1000	0.35	—	—
22	1100	0.29	—	—
24	1200	0.25	0.25	0.25

U_d Direct voltage (100%).
$\nu = 2$ to 24 Ordinal number of superimposed alternating voltage component.

The pulse number of the thyristor circuit determines the ordinal number of the alternating voltage components occurring. U_{vi}/U_{di} is dependent only on the ordinal number respectively frequency, but not on the pulse number p of the rectifier. For example, the same amplitude of the ideal alternating voltage:

$$\frac{U_{vi}}{U_{di}} = 0.99\%$$

is present with all three pulse numbers when the ordinal number $\nu = 12$ ($f = 600$ Hz).

The effect of raising the pulse number p is favourable if thereby the components with the highest voltage come out with a lower ordinal number respectively lower frequency.

The alternating voltage content actually occurring in operation at the d.c. connections of the rectifier set is given by:

$$w = \frac{U_{s\alpha}}{U_{d\alpha}}.$$

2.3.2 Filter sections

Filter sections (low-pass filters) are necessary with rectifiers for maintaining the required purity of the d.c. output voltage. Figure 2.5(a) shows a typical filter section. Figure 2.5(b) shows the attenuation curve. At a low frequency the attenuation is very slight (pass band). As the frequency becomes higher the attenuation increases accordingly (stop band).

With surges in load the storage effect of a filter section, particularly with large capacitances, permits within certain limits the attenuation of fluctuations in the d.c. output voltage.

Figure 2.6 shows filter sections as they are to be found in thyristor-controlled rectifiers. Sections designed for single-phase systems are normally fitted with one low-pass filter (Fig. 2.6a). As higher outputs can normally be taken from rectifiers designed for three-phase systems e.g. two filter sections are employed (Fig. 2.6b).

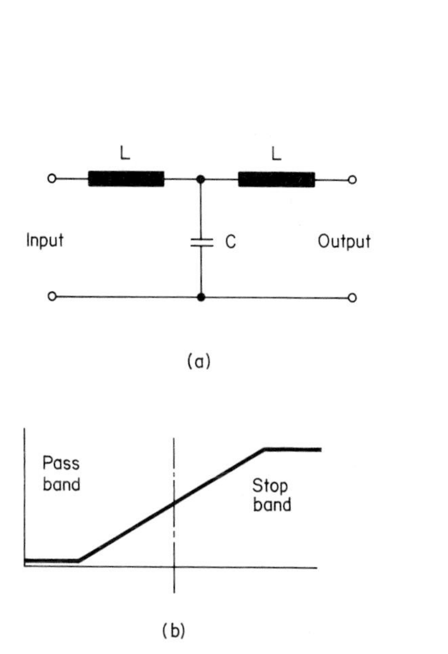

Fig. 2.5
Filter section (a) with attenuation curve (b)

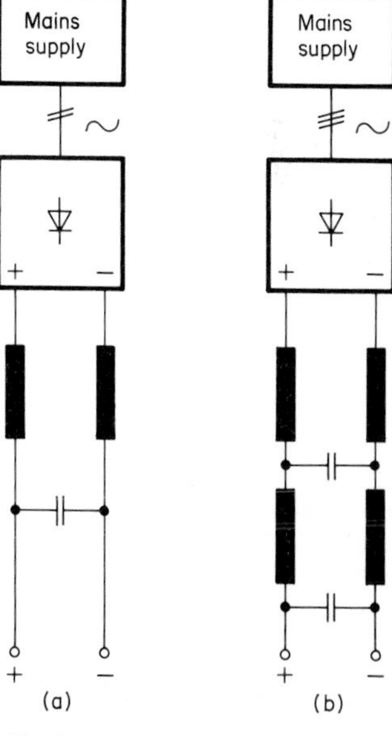

Fig. 2.6
Filter sections with thyristor-controlled rectifiers

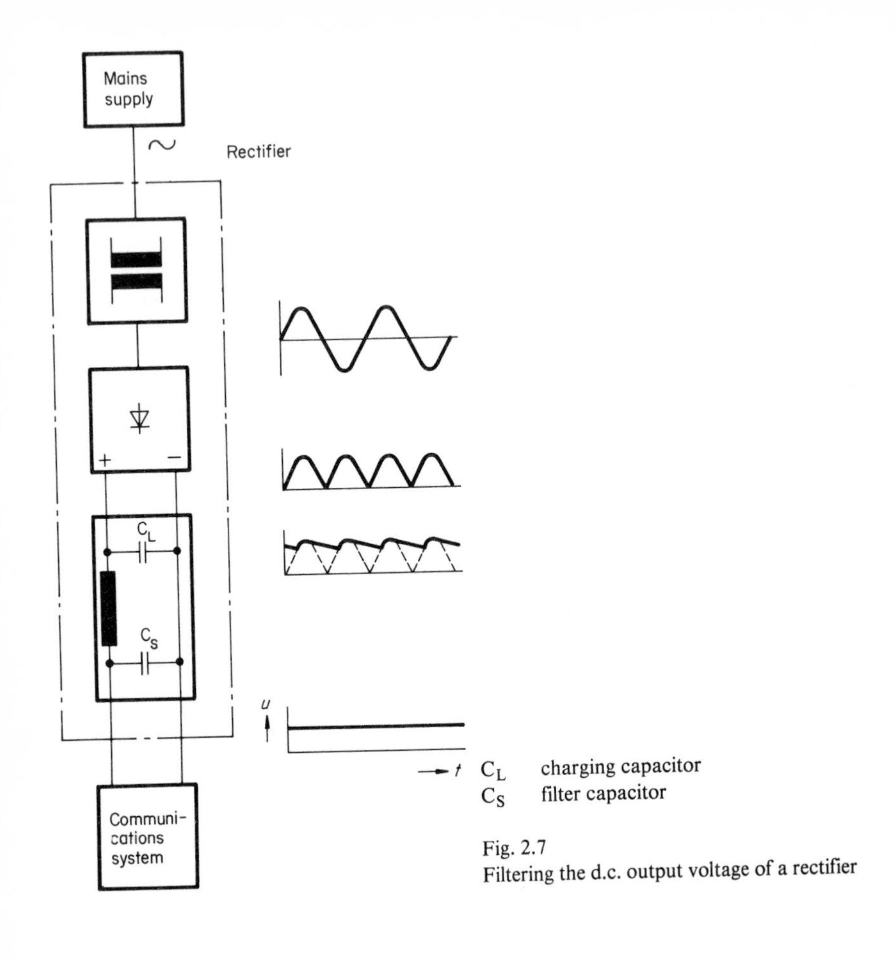

Fig. 2.7
Filtering the d.c. output voltage of a rectifier

The effect of filtering can be seen with the aid of the simplified circuit shown in Fig. 2.7.

There is a (stepped-down) alternating voltage at the transformer output. Parallel with the output of the rectifier circuit is the charging capacitor C_L which acts as an energy store. The process of charging and discharging C_L is shown in Fig. 2.8.

The amplitude of the discharge current and the size of the capacitor influence the level of the direct voltage. If there is a small load at the output, the charging capacitor is scarcely discharged and the direct voltage behaves as in Fig. 2.8, curve 1. With a small load the ripple is slight. With a large load at the output an (equally large) capacitor gives off a large part of the stored energy which is shown by curve 2 in Fig. 2.8. It can be seen that the ripple has increased.

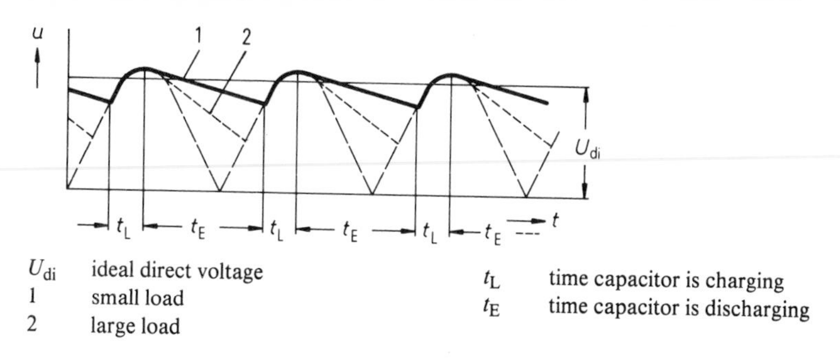

U_{di}	ideal direct voltage	t_L	time capacitor is charging
1	small load	t_E	time capacitor is discharging
2	large load		

Fig. 2.8 Effect of the charging capacitor

2.3.3 Reducing the interference voltage by the type of wiring installation

Sources of interference voltages within the communications system also have to be taken into account in addition to those in the power supply units. These interference voltages originate from selector drives, relay contacts, coupling fields, etc., and produce a broad interference voltage spectrum.

Attempts have been made to reduce the effects of these sources of interference by means of additional capacitors in the frames. This is not completely successful, however, so the interference voltage spectrum is largely eliminated using the filtering means present in the power supply system. The communications system must therefore be connected to the power supply system across as low an impedance as possible.

It is also important that the lines and bars are arranged and laid with low induction.

When installing a power supply system, the length of the lines is predetermined by spatial considerations and the copper cross-section is determined by, among other things, the specified voltage drop. The inductance can only be influenced by the separation distance between the lines. To illustrate the influence which the bar layout has on impedance, and thus on the interference voltage, Fig. 2.9 compares some examples. A line length of 20 m (two-wire line) and a cross-section in the case of 1000 mm^2 is assumed as standard. Example 1 relates the interference voltage, assuming 100%, to an impedance of 100 mΩ. The diagram clearly shows the benefit achieved by the close parallel exposure in the fourth example.

When running cables in parallel the positive and negative conductors of a circuit must be laid close together, as with flat copper bars.

An arrangement of the cables as in Fig. 2.10(a) is unsuitable (largest impedance and hence highest interference voltage). The best values are obtained with the arrangement shown in Fig. 2.10(b), but this is difficult to achieve because of the many cable

Example	Type of laying (distances in mm)	Inductance in µH about	Impedance at 800 Hz in mΩ about
1	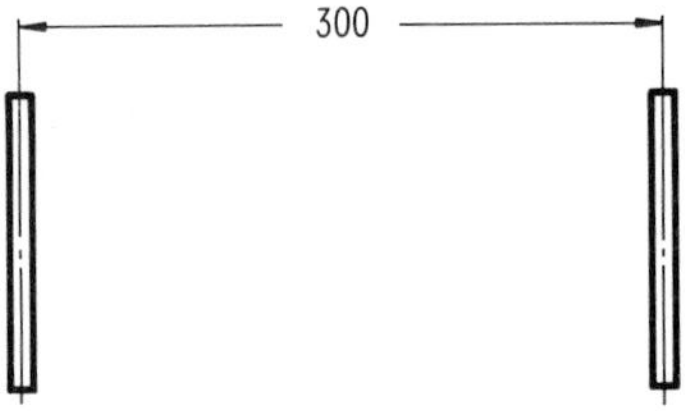 300	20	100
2	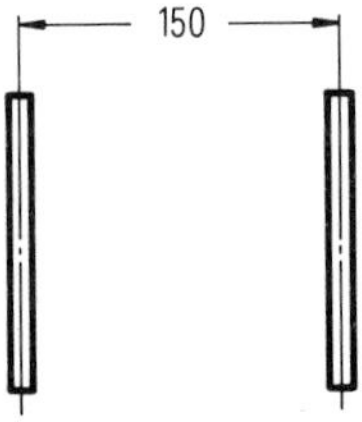 150	15	75
3	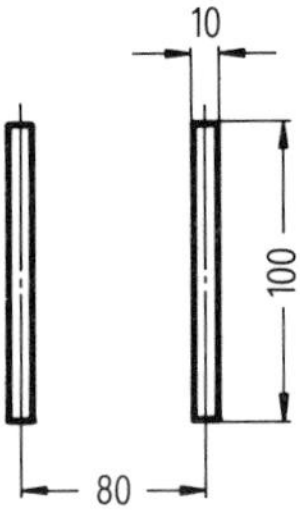 10; 100; 80	10	50
4	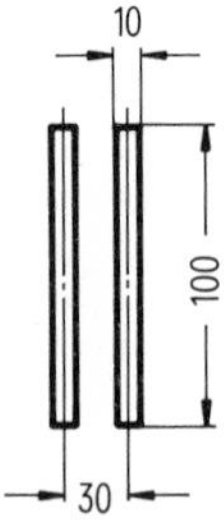 10; 100; 30	5	25

Fig. 2.9 Influence on impedance of how the bars are laid

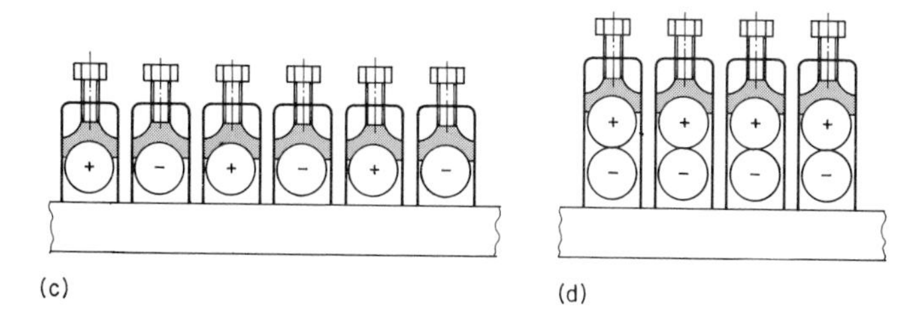

Fig. 2.10 Arrangement of direct current cables

transpositions that then arise. For this reason the choice in practice falls upon the arrangement as in either Fig. 2.10(c) or (d).

2.4 Degree of Radio Interference and Limit Classes

Interference voltages can be transmitted along lines as through the air by the propagation of radiowaves. This interference can be registered acoustically in the case of radio reception and in the form of stripes, etc., in the case of television reception. For this reason all electrical equipment, including power converter, must be *suppressed.*

Radio interference is reduced as much as necessary by suppression using capacitors and chokes. Degrees of radio interference are distinguished according to the level of the remaining radio interference voltage.

The *degree of radio interference*[1]) depends on the frequency of the radio interference (continuous and crackling interference). In power supply systems rectifiers, compensators, etc., with frequencies of less than 10 kHz, which can inadvertently generate high-frequency interference, are suppressed at least in accordance with the degree of radio interference N (Fig. 2.11), where N stands for the normal degree of interference and applies to residential areas (cf. also Table 2.1, line 7).

[1]) The degrees of radio interference respectively the limits of the radio interference voltage for continuous interference are contained in VDE 0875/DIN 57875.

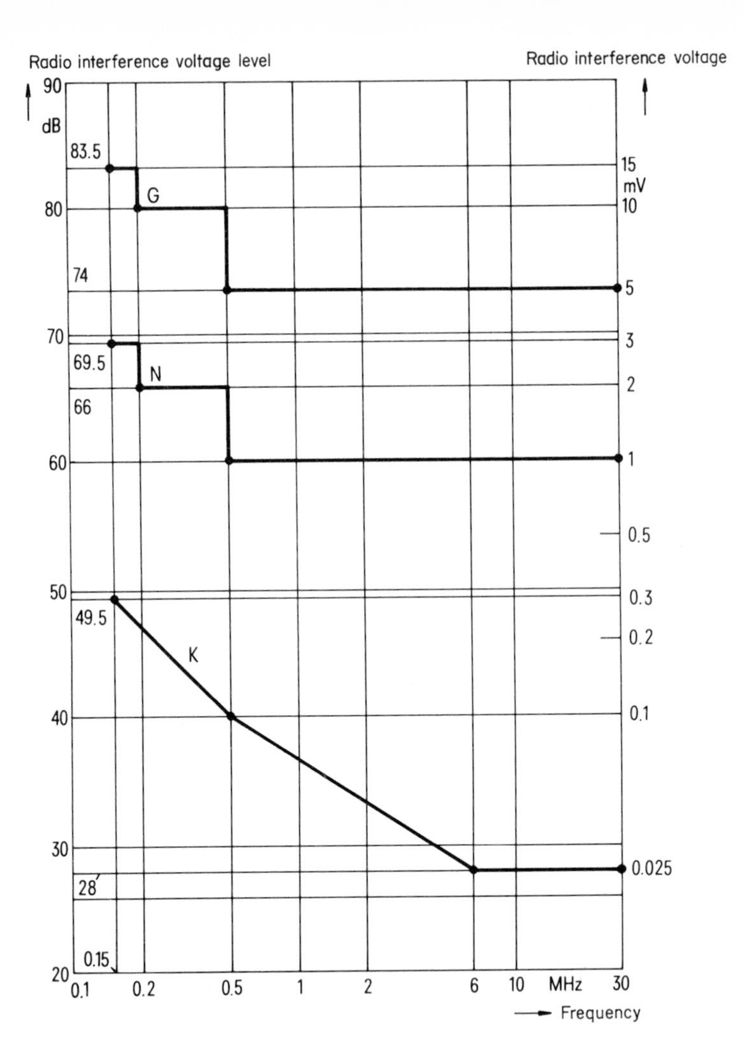

Fig. 2.11
Limits for radio interference voltage for continuous interference for degrees of radio interference K, N and G to VDE 0875/DIN 57875

Apart from the degree of radio interference N there is also:

- ▷ a small degree of interference K for radio stations,
- ▷ a coarse degree of interference G for industrial areas,
- ▷ a degree of radio interference 0 with freedom from interference.

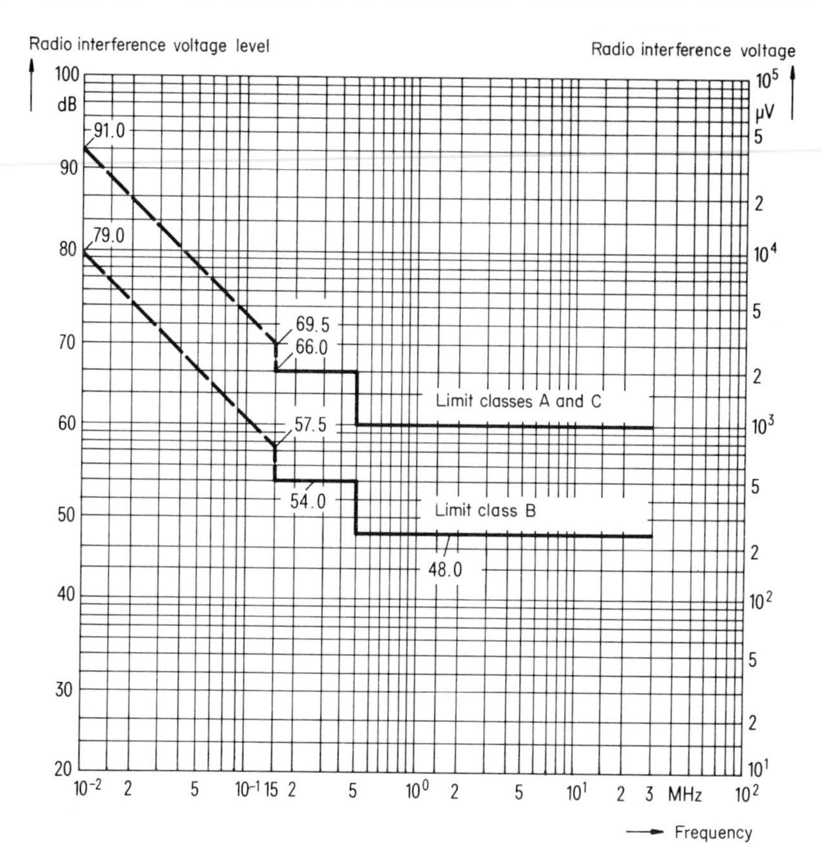

Fig. 2.12
Limits for radio interference voltage with equipment causing interference in a range above 10 kHz (to VDE 0871/DIN 57871)

With a single rectifier the degree of radio interference N can be indicated and adhered to. If, however, a number of rectifiers are placed in parallel, it is only with difficulty that the overall degree of radio interference can be calculated in advance. The said conditions therefore are normally tightened up on the direct current side to, for example, N – 12 dB.

The *limit class* designates the assignment of high-frequency equipment and systems to different limit values. In the case of telecommunications power supply equipment causing continuous interference of more than 10 kHz, VDE 0871/DIN 57871 (Fig. 2.12) permits three limit classes A, B and C. Normally limit class B is taken as the basis when designing equipment.

2.5 Current Distribution System and Voltage Drops

The influence upon the current distribution system through interference voltages from the power supply system, the reaction of equipment to the current distribution system, plus the mutual influencing of equipment, must all be kept to a minimum.

A current distribution system normally comprises:

power supply lines,
earth lines,
protective lines,
cutout devices (fuses),
diodes and capacitors for decoupling line sections.

Figures 2.13 and 2.14 examine the voltage drops of individual communications systems. The voltage drop occurring on the supply lines between the telecommunications power supply system and the equipment cannot be neglected. This applies in particular in the case of power failures with subsequent battery discharge.

The permissible operating voltage for systems below the voltage drops have been laid down and allocated to the individual sections of the current distribution system (cf. Table 2.1, line 8). If a compensator is used, the permissible voltage drop in the current distribution system can be raised by about 1 V. This leads to smaller wire cross-sections which at least compensates in part for the costs of the compensator.

The continuously permissible tolerance for the operating voltage (cf. Table 2.1, line 3) for the ESK 10 000 E system is 44 to 53 V. If it is required to discharge the battery only to 1.79 V/cell and 26 cells are taken as the basis, the maximum overall voltage drop (Fig. 2.13) is calculated as follows:

$$26 \times 1.79\ \text{V} \approx 46.6\ \text{V}; \quad 46.6\ \text{V} - 44\ \text{V} = 2.6\ \text{V}.$$

This voltage drop between the poles of the battery and the equipment in the last rack row is broken down as follows:

Fuses, switches, etc., in the power supply system	0.4 V
Battery and load lines	1.0 V
Rack row fuses	0.1 V
Branch load line in the rack rows	0.2 V
Racks (individual fuses, etc.)	0.9 V
Total voltage drop	2.6 V

The figures in brackets in Fig. 2.13 relate to a 60 V communications system (e.g. EMD). An overall voltage drop of max. 2.4 V occurs here. This system of current

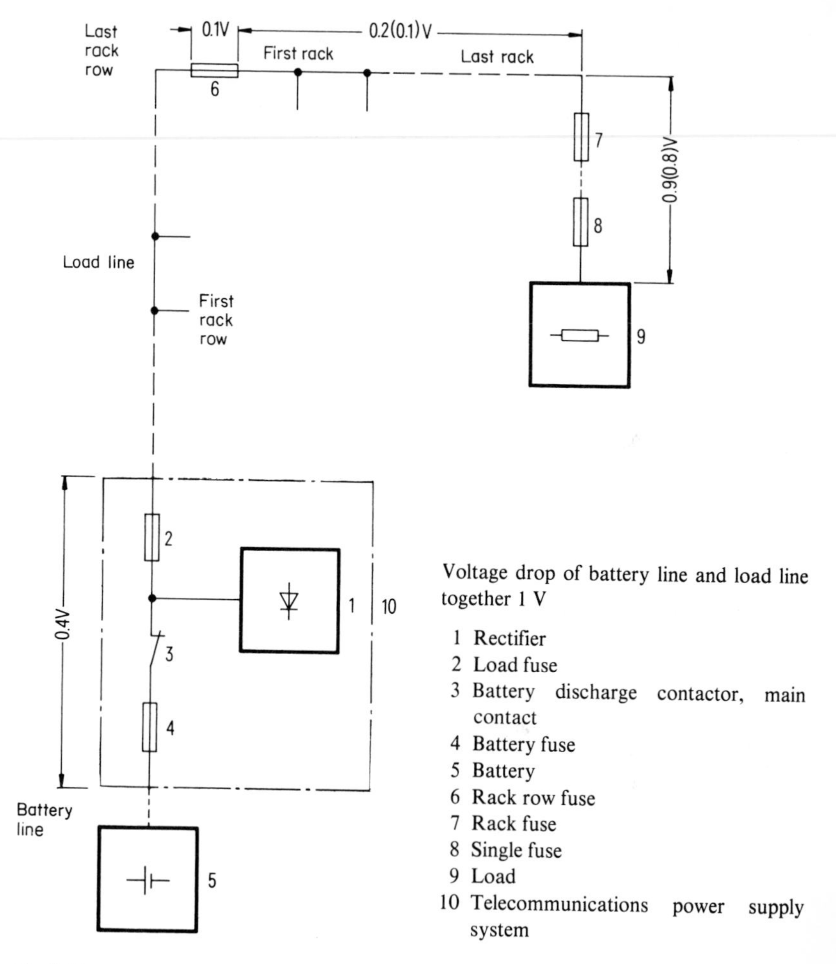

Voltage drop of battery line and load line together 1 V

1 Rectifier
2 Load fuse
3 Battery discharge contactor, main contact
4 Battery fuse
5 Battery
6 Rack row fuse
7 Rack fuse
8 Single fuse
9 Load
10 Telecommunications power supply system

Fig. 2.13
Current distribution system and voltage drops in a 48 V communications system, e.g. ESK 10 000 E, with 26-cell lead battery; total voltage drop max. 2.6 V. Figures in brackets relate to a 60 V communications system, e.g. EMD with 31-cell lead battery; total voltage drop max. 2.4 V

distribution is used primarily in non-European countries (for EMD systems) together with 31-cell lead batteries without a compensator.

As a further example Fig. 2.14 shows an outline of the voltage drops occurring with the EWSD system. The maximum total voltage drop here is 1.8 V, with 25 cells taken as the basis.

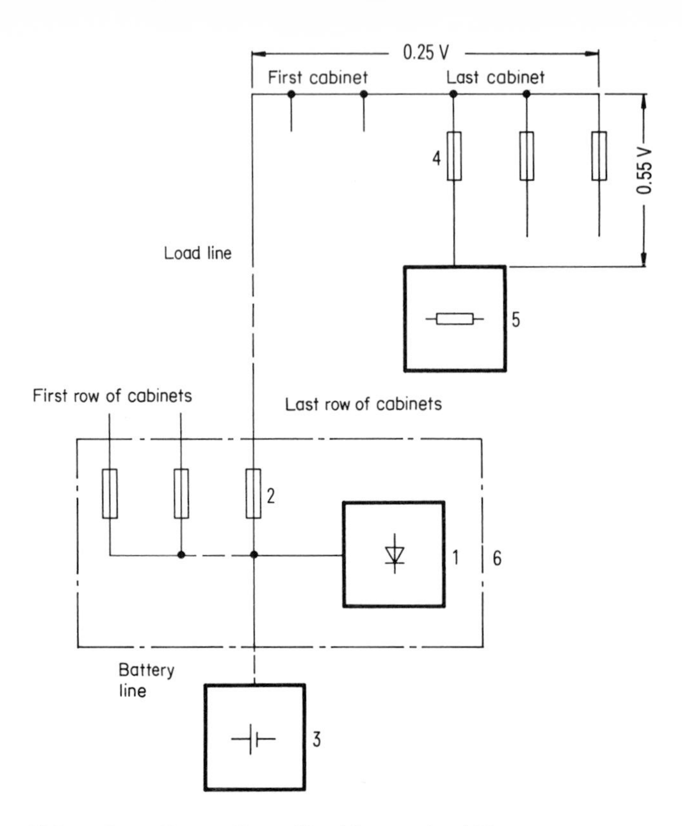

Voltage drop of battery line and load line together 1 V

1 Rectifier
2 Load fuse
3 Battery
4 Cabinet fuses
5 Load
6 Telecommunications power supply system

Fig. 2.14
Current distribution system and voltage drops in a 48 V communications system, e.g. EWSD, with 25-cell lead battery; total voltage drop max. 1.8 V

2.6 Availability of the Power Supply

Availability A is understood essentially to be the *reliability* of the telecommunications power supply. Availability with present-day systems is very high. It can be shown as follows:

$$A = \frac{MTBF}{MTBF + MTTR}$$

where

$MTBF$	is the mean time between failures in years and
$MTTR$	is the mean time to repair in minutes.

Non-availability N is represented by:

$$N = \frac{MTTR}{MTBF + MTTR}.$$

The values shown in Table 2.3 were observed in 8000 direct current power supply systems operated in West Germany.

The public mains usually serve as the primary source of energy for a telecommunications power supply system. Although the reliability of the mains in West Germany can be regarded as relatively high, reliable operation of communications systems is not possible without the provision of alternative power sources (battery and/or standby power supply system).

In case of a power failure the alternative power supply source takes on the task of feeding the communications system. In parallel mode this is possible without additional switching equipment. Changeover mode requires that the connecting devices and subassemblies involved in the switching operation ensure the greatest reliability.

The capacity and the number of cells chosen for a battery depends on the time it is required to supply energy, the magnitude of the discharge current and the permissible bottom voltage of the communications system. Thus it is possible to have an alternative power supply available for a limited time. If a supply is to be guaranteed to be available without any limit on time, an alternative to the mains will

Table 2.3
Reliability of direct current power supply systems

Number of total failures through faults, yearly	20
$MTBF$	400
$MTTR$	88
Availability A	0.9999996
Non-availability N	0.4×10^{-6}

be necessary. In this case the capacity of the battery can be made smaller, as this is only necessary for switching to the mains equivalent mode without interruption. As a reserve of energy a battery is normally adequate for 6 hours. If a fixed standby power supply system is provided, the energy reserve of the battery can be reduced to 4 hours.

3 Operating Modes of a Direct Current Power Supply System

The description of the operating modes of direct current power supply systems (Fig. 3.1) essentially relates to the telecommunications power supply systems in sections 3.1 to 3.6. The most important terms and names for the modes are defined in VDE 0800 respectively 0510.

Table 2.1 applies to the operating voltages of communications systems.

3.1 Battery Mode

In the battery mode ('charge-discharge mode') two batteries are always necessary, as in the earlier layout (cf. Fig. 1.7). While one battery is supplying the communications system (discharging), the second battery is being charged. At certain intervals the functions of the batteries are reversed.

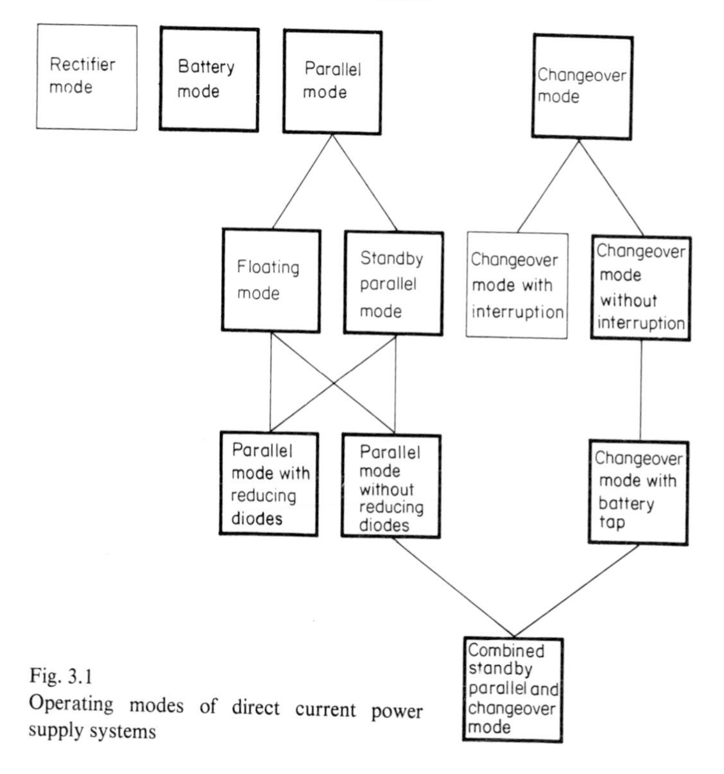

Fig. 3.1
Operating modes of direct current power supply systems

This mode, because of its relatively low efficiency and the particularly high demand on the battery, is only used nowadays in telecommunications power supply systems when the mains supply fails and the supply of alternating current must be guaranteed with a power supply system independent of the mains. For example, two standby units each with a relatively short operating time charge up the battery. The units are only in operation for a few hours, while the load is supplied continuously from the battery. This enables longer maintenance intervals to be achieved for the units.

3.2 Rectifier Mode

In the rectifier mode, also called the direct feed mode, there is *no* battery. The communications system is supplied with direct voltage directly from the mains via the rectifier (Fig. 3.2).

The supply is interrupted for the duration of any power failure or in the event of a breakdown of the rectifier. The rectifier automatically switches on again on return of the mains.

This mode is used with small to medium-sized communications systems when interruptions in operation can be accepted.

In the case of PABX systems it is possible to switch the exchange line. This means that in the event of a power failure exchange line calls can continue to be made from one or more extensions, though direct calls from extension to extension are not possible.

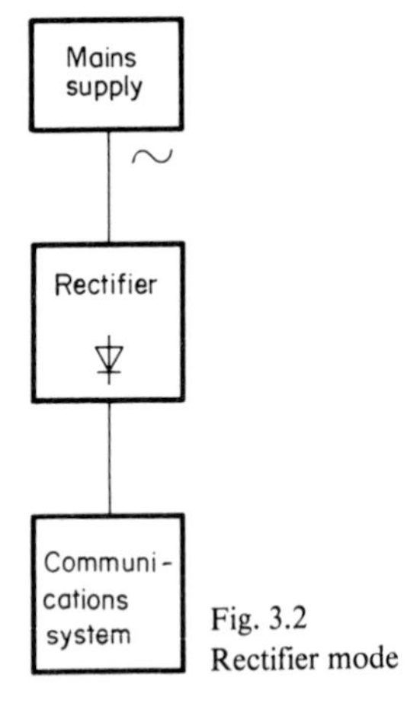

Fig. 3.2
Rectifier mode

3.3 Parallel Mode

If the communications system is required to continue operation during a power failure, or in the case of other troubles, a reserve of energy (preferably in the form of a lead battery) should be kept ready. In the parallel mode the rectifier, battery and communications system are constantly connected in parallel (Fig. 3.3). If the rectifier fails, the battery takes over the further supply of the communications system until the rectifier, on return of the mains, starts operating again. The rectifier then supplies the communications system again and also charges the battery.

The advantages of parallel mode are thus: the maintenance of an uninterrupted supply to the communications system without additional switching equipment and the absorption of surges in load by the battery.

To avoid the dependence on battery capacity and line length as regards filtering all rectifiers are smoothed and stabilized for the interference voltage values required by the communications system, regardless of the mode in which they are used (cf. Table 2.1).

A distinction is made in the parallel mode between the *floating mode* and *standby parallel mode.*

3.3.1 Floating mode

In the floating mode the rectifier can handle the communications system's normal energy requirement, though it cannot deal with the peak current. In this case the battery provides the current over and above the rectifier's rated current

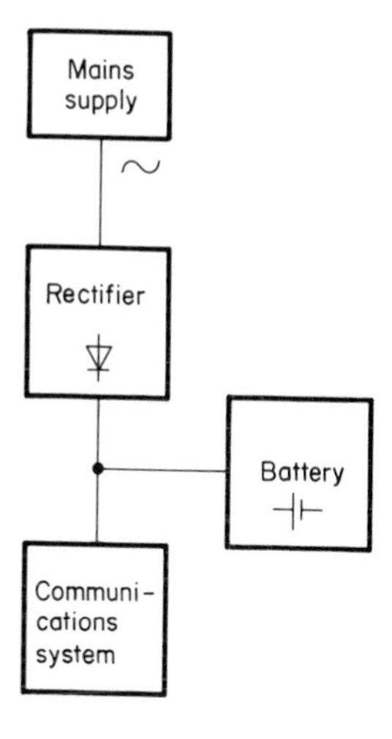

Fig. 3.3
Parallel mode

($I_{load} > I_{rated}$). If the energy requirement (outside busy hours) becomes less again, the battery takes a charging or trickle charging current from the rectifier. It is thus alternately used to supply additional current or is charged. It can happen in the floating mode that the full battery capacity is not available. This means a shorter reserve time for bridging a power failure, unless a correspondingly larger battery capacity is chosen with regard to a certain reserve time. The life of the battery is shortened in the floating mode which is why today it is usually the second variant of the parallel mode that is used, viz. standby parallel mode.

3.3.2 Standby parallel mode

In the standby parallel mode the rectifier always covers the communications system's whole energy requirement. The battery is also supplied with a 'trickle charge' by the rectifier. It is therefore available with its full capacity in case of a power failure or breakdown of the rectifier, provided the interval of time between the previous duty and renewed discharging has been sufficient to charge the battery.

If in an exceptional case the communications system suddenly needs a current larger than the rated current of the rectifier, then the battery supplies it. The standby parallel mode then temporarily changes to the floating mode.

The advantages of the standby parallel mode compared with the floating mode are thus that:

▷ a longer battery life due to the continuous trickle charging is available and
▷ the calculated capacity of the battery can always be assumed as, in normal operation, it is always fully charged.

If a certain number of battery cells are provided for a communications system according to its minimum permissible operating voltage, which is reached when the battery finishes discharging but below which it must not fall (system-conditioned final voltage $U_{s\,min}$), the resultant supply voltage can become too high for the communications system if it is connected in parallel. This applies in particular to conventional systems.

In present-day communications systems a voltage swing of such a size is taken into account so that they harmonize with the battery voltages.

The parallel mode can be further divided into two subgroups taking into consideration these aspects:

▷ *parallel mode with reducing diodes* (cf. Fig. 3.4) and
▷ *parallel mode without reducing diodes* (cf. Fig. 3.7).

The operating states of these two modes is explained below.

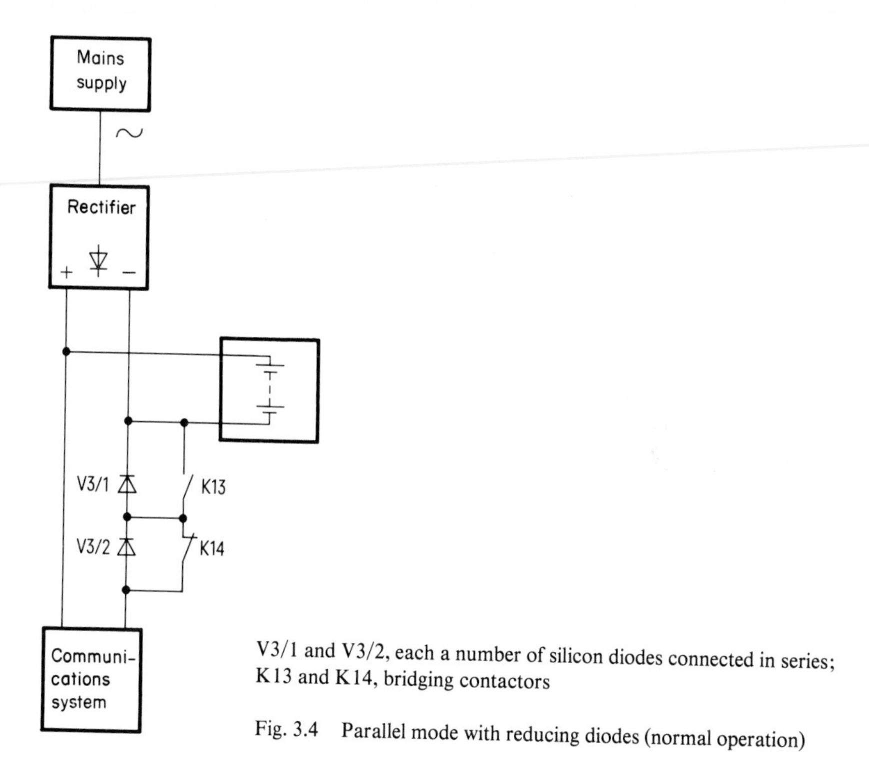

V3/1 and V3/2, each a number of silicon diodes connected in series; K13 and K14, bridging contactors

Fig. 3.4 Parallel mode with reducing diodes (normal operation)

Parallel mode with reducing diodes

Normal operation-supply to the communications system and trickle charging of the battery with 2.23 V/cell: In normal operation the rectifier supplies the communications system. The voltage for a trickle charge of 2.23 V/cell is applied to the battery (Fig. 3.4). As the resultant voltage (2.23 V × number of cells) is too high for the communications system, it is reduced to the desired value by reducing diodes (counter cells). The voltage drop (in the forward direction) of silicon diodes is used for this. The bridging contact K13 is thereby opened with the group of reducing diodes V3/1 thus becoming effective. The group of reducing diodes V3/2 is bridged by the bridging contactor K14 and is thus ineffective.

Power failure-battery discharge (communications system supplied from the battery): In the event of a power failure the battery takes over the provision of a power supply to the communications system without interruption. When the battery voltage falls to a certain value (cf. Fig. 3.5) the group of reducing diodes V3/1 are voltage-dependently bridged by the bridging contactor K13 so that the whole battery voltage is now available for supplying the communications system.

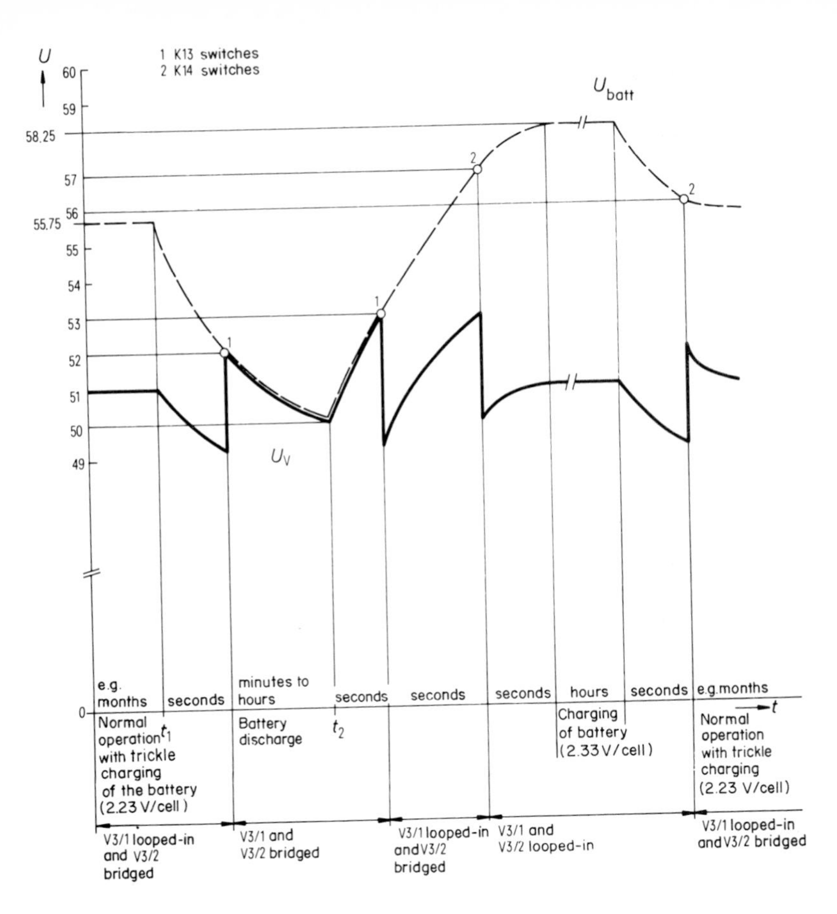

t_1 power failure t_2 return of mains

Fig. 3.5
Operating conditions – behaviour of load and battery voltage (with 25-cell lead battery)

Mains return-supply of the communications system and charging of the battery with 2.33 V/cell: On return of the mains the battery is supplied with 2.33 V/cell for rapid recharging (hour range). For this purpose a further group of reducing diodes (V3/2) is looped in addition to the V3/1 group. Both bridging contactors K13 and K14 are opened and therefore both groups of reducing diodes V3/1 and V3/2 are effectively connected. In this way the supply voltage for the communications system is kept within the permissible limits, even when in this regime.

At the end of charging there is a switch back to normal operation with trickle charging.

Fig. 3.6
Controlled rectifier with thyristor power section 60 V/100 A

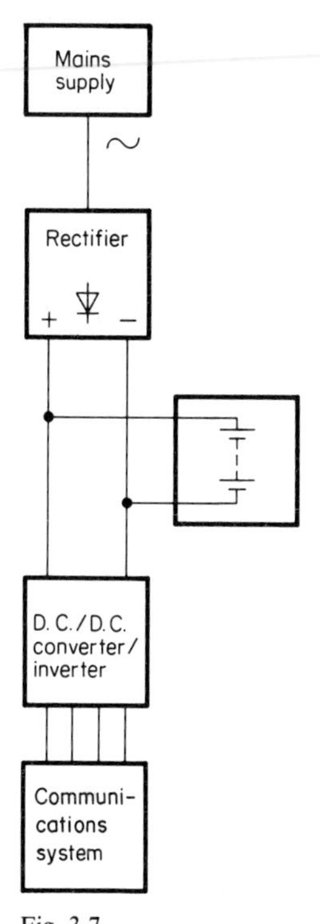

Fig. 3.7
Parallel mode without reducing diodes

Figure 3.5 shows the behaviour of load and battery voltages in the various operating conditions and the voltage-dependent looping in and bridging of the groups of reducing diodes. The voltage levels shown for the looping-in of the group of reducing diodes V3/2 and bridging have recently been changed from 57 to 58 V and from 56 to 57 V, respectively.

Figure 3.6 illustrates a design example of a rectifier for parallel mode with reducing diodes.

Parallel mode without reducing diodes

In present-day communications systems d.c./d.c. converters and inverters whose permissible input voltage range is so great that they can be directly connected in parallel with the battery (Fig. 3.7) are increasingly used.

If the mains fail, here too the battery takes on supply of the communications system without interruption.

3.4 Changeover Mode

In normal operation one rectifier supplies the communications system and a second one supplies the battery (Fig. 3.8). The battery is only connected with the communications system in the event of a power failure. It then continues supplying the system.

A distinction is made between (cf. Fig. 3.1):

changeover mode *with* interruption,
changeover mode *without* interruption.

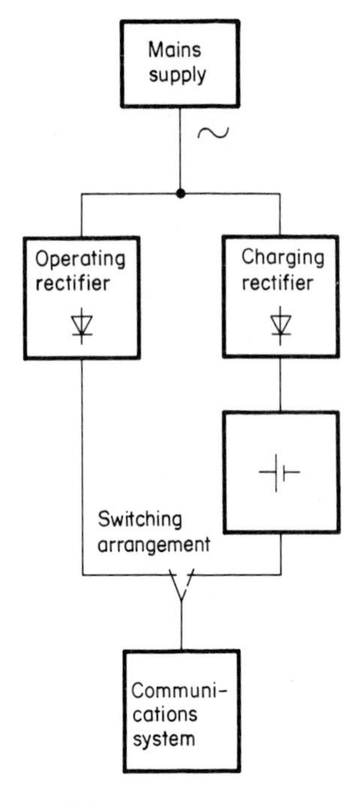

Fig. 3.8
Changeover mode

Changeover mode with interruption

Changeover mode *with* interruption means that the supply to the communications system is interrupted *for a short time* when changing over to battery operation and also when changing back to mains operation. As this is usually not permissible, this mode is rarely used in practice.

Changeover mode without interruption

For the above-mentioned reasons changeover mode *without* interruption is used in telecommunications power supply systems.

3.4.1 Changeover mode with battery tap

'The changeover mode with battery tap' is a variant of the changeover mode without interruption. It is used in power supply systems for communications systems with a rated current from about 100 to 10 000 A.

When the changeover mode is discussed later it is to be understood that this means the changeover mode without interruption with battery tap. Instances where this is not so will be specially indicated.

The operating conditions respectively operating state are shown in Fig. 3.9.

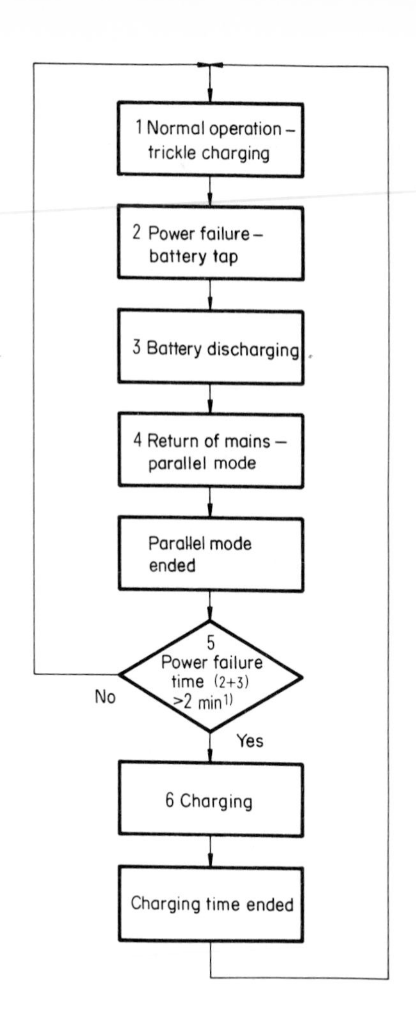

Fig. 3.9
Operating state for changeover mode with battery tap

[1]) 2 min – applies to 60 V systems; 3 min for 48 V ones

1 Normal operation-supply to the communications system and trickle charging of the battery with 2.23 V/cell: Here, with the mains available, the system is supplied by rectifier 1 (operating rectifier, Fig. 3.10), while at the same time the trickle charging voltage is applied to the battery by rectifier 2 (charging rectifier). Thus only so much current flows across the battery that its inherent discharging is covered and the battery thereby remains fully charged.

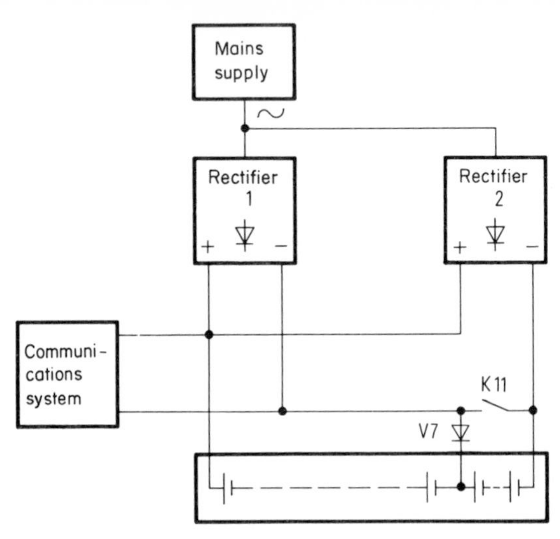

V7 Tapping diode
K11 Battery discharge contactor

Rectifier 1 Operating rectifier
Rectifier 2 Charging rectifier

Fig. 3.10 Changeover mode with battery tap (normal operation)

2 Power failure – transfer to battery discharging with aid of battery tap: If the mains voltage fails or the supply of energy from the rectifier is absent due to some other trouble, the communications system is supplied via tapping diode V7 from a certain number of the battery's cells until battery discharge contactor K11 closes.

With the battery tap (Table 3.1) it is possible to transfer to the battery operation *without interruption.*

It is characteristic that tapping diode V7 is *only* conducting during the time the battery discharge contactor is changing over from normal operation to the battery operation. In all other operating conditions it is polarized in the reverse direction.

The position of the battery tap is selected so that even when charging, for example, 26 battery cells with 2.33 V/cell no higher voltage occurs than d.c. output voltage of rectifier 1. There would otherwise be a passage of current from the battery to the communications system via the tapping diode.

With almost all systems there is, in the case of power failure, an *immediate* switch to battery discharging via the tapping diode. The time during which current passes through the tapping diode is only some 100 ms.

An exception is 48 V systems with a 26-cell battery and tapping at the 22nd cell. In these systems the switch is voltage-dependent, i.e. only when the voltage has reached a certain minimum is the discharge contactor released, thereby changing over to battery discharging (all cells). This prevents a possibly too high battery voltage reaching the system.

Table 3.1 List of battery tap positions and cell numbers

System	Battery tap at
With 25-cell lead battery	21st cell
With 26-cell lead battery	22nd cell
With 30-cell lead battery	26th cell
With 31-cell lead battery	26th cell

3 Battery discharging: After the battery discharge contactor K11 has closed, the communications system is supplied from the whole battery (all cells). During a power failure the voltage of the communications system is the same as the battery voltage, i.e. the supply voltage falls in accordance with the battery discharge characteristic as a function of the size of the load and the time that discharging lasts.

If necessary, a compensator can also be used (only for 60 V systems of the Federal German Postal Administration, cf. Section 3.5).

4 Mains return – parallel mode with battery: The rectifiers automatically switch on after the mains voltage returns. For a certain, adjustable transition time (e.g. 30 min to 3 h) rectifiers 1 and 2 are connected in parallel with the battery and communications system, thereby ensuring that the rectifiers are already supplying energy before the battery is separated.

If the rectifiers are not used to supply energy in the parallel mode (with a low load and after a power failure of short duration the battery voltage can be greater than the d.c. output voltage from the rectifiers), there can be a shock-like loading of the rectifiers after this operating condition comes to an end, which could result in an impermissible voltage variance and, with it, *switching off* of the rectifiers.

Measures taken to prevent this are:

▷ to raise the voltage of the rectifiers during the parallel mode,
▷ to extend the time setting of the 'parallel mode time' relay.

5 Power failure time decision: If the *duration of the power failure* (2 + 3) has been *shorter than 2 min* (60 V systems) *respectively shorter than 3 min* (48 V systems) and the parallel mode time is at an end, there is a switch back to *normal operation* (see 1). The contact of the battery discharge contactor K11 breaks and the battery is thereby separated again from the communications system.

6 Charging: If the *duration of the power failure* (2 + 3) has been *longer than 2 min* (60 V systems) *respectively longer than 3 min* (48 V systems) and the parallel mode time is at an end, there is a switch to *battery charging* (2.33 V/cell).

Here, too, once the parallel mode time has expired, the contact of the battery discharge contactor K11 breaks and the battery is thereby separated from the communications system. Rectifier 1, as in the parallel mode (see 4), continues supplying the communications system. Rectifier 2 takes on charging the battery with 2.33 V/cell. After an adjustable time (e.g. up to 24 h) rectifier 2 switches back to trickle charging (2.23 V/cell, see 1).

Initial charging: The battery can be charged up to 2.7 V/cell for the start-up charging of uncharged batteries or the subsequent special treatment of a battery after damage (special charging). In this case, e.g. in systems with a battery switching panel, the appropriate battery switch is set to initial charging by hand.

One of the rectifiers is also switched from the load bar to the charging bar to initial charging. Another term used for this charging operation is 'forming charging'.

3.5 Combined Standby Parallel and Changeover Mode

The 'combined standby parallel and changeover mode' (Figs 3.11 and 3.12) represents a mixture of the parallel mode without reducing diodes (Fig. 3.7) and the changeover mode (Fig. 3.10).

Compared with Fig. 3.10 there is also a load 2 in Fig. 3.12 in addition to load 1. Load 1 represents telecommunications equipment requiring a supply voltage with narrow tolerances. The changeover mode is therefore used here. Rectifier 1 normally supplies load 1 with constant direct voltage.

A compensator is inserted in the supply line so that the supply voltage remains constant within the tolerance range, even in the case of a power failure. The compensator (supplied from the battery) supplies an additional voltage of up to 7 V in the case of a power failure. Bridging contactor K2 is then opened. The smaller the battery voltage becomes in the course of its discharging, the larger becomes the compensator's additional voltage.

The voltage that is fed to the battery, or is given off by it, is always applied to load 2. By load 2 is meant where wide tolerances are permitted in the supply voltage, e.g. the d.c./d.c. converter and inverter. For this reason the standby parallel mode is used here.

Compared with Fig. 3.10, Fig. 3.12 also shows the tapping diode V7 and in addition the decoupling diode V6. This diode V6 prevents rectifier 1 from participating in the charging of the battery in the operating state of parallel mode.

The operating state of the combined standby parallel and changeover mode is described below.

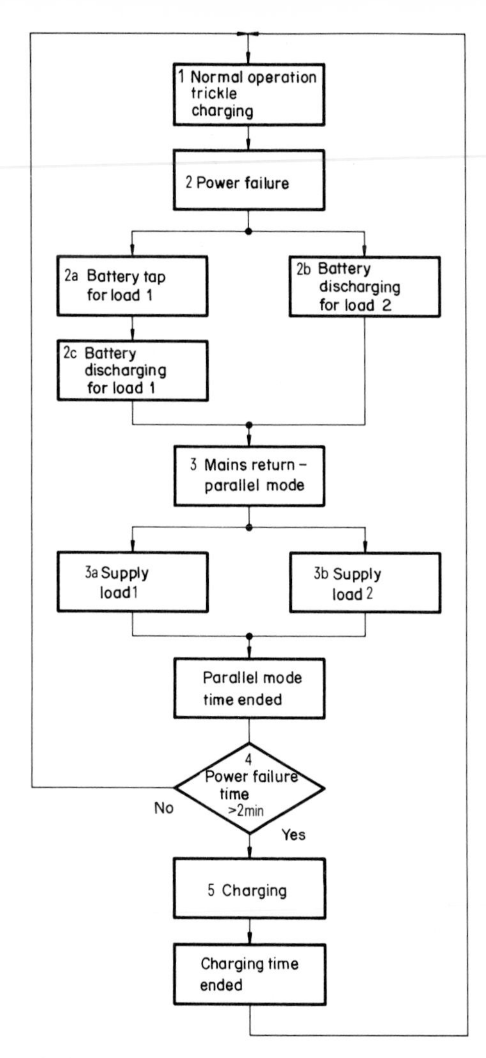

Fig. 3.11
Operating state for combined standby parallel and changeover mode

1 Normal operation – supply to the communications system and trickle charging of the battery (2.23 V/cell): If the mains voltage is available, rectifier 1 supplies the communications system (load 1) with constant direct voltage via the compensator which is bridged in this operating condition. Rectifier 2 supplies the trickle charge voltage of 2.23 V/cell for the battery. Parallel with the battery is the communications system (load 2), which thereby receives the same supply voltage as the battery.

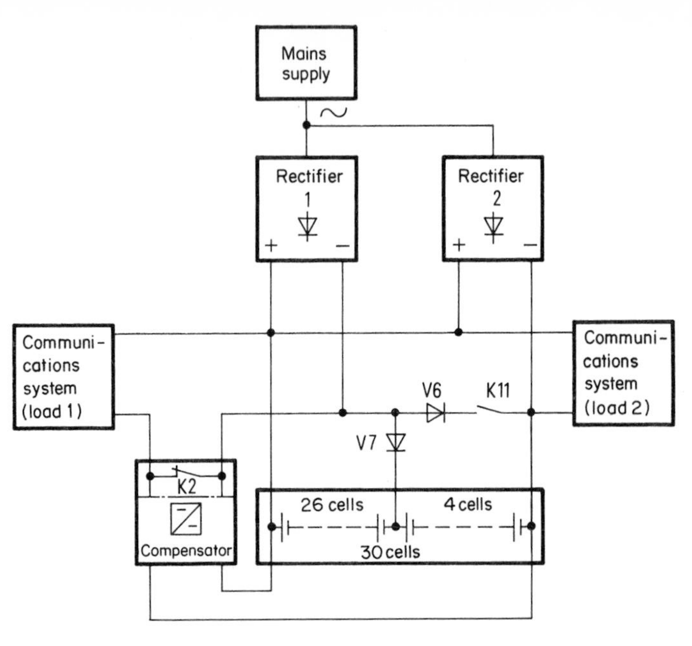

V6	Decoupling diode	Rectifier 1 Operating rectifier
V7	Tapping diode	Rectifier 2 Operating rectifier and charging rectifier
K2	Compensator's bridging contactor	
K11	Battery discharge contactor	

Fig. 3.12
Combined standby parallel and changeover mode (normal operation)

2 Power failure

2a Battery tap – transition to battery discharging with aid of battery tap, supply to communications system (load 1): Figure 3.12 illustrates that in the event of a power failure load 1 is supplied without interruption by the voltage from the 26 battery cells, via the tapping diode V7, until the battery protective discharge contact K11 closes. Changeover lasts some 100 ms.

2b Battery discharging – supply to load 2: The voltage applied to load 2 during the power failure is the same as the battery voltage. It is to be borne in mind here that this voltage drops in accordance with the battery's discharging characteristic depending on the size of the load and the length of time the battery is discharging.

2c Battery discharging – supply to load 1: Once the battery's discharge contactor K11 has closed load 1 is supplied from all the battery cells via decoupling diode V6.

If the battery voltage falls below a certain value, bridging contact K2 breaks in the compensator. By means of the additional voltage from the compensator the supply voltage for load 1 is kept within the tolerance range at about that voltage level corresponding to normal operation.

3 Mains return – parallel mode: After return on the mains rectifiers 1 and 2 switch on again with the same d.c. output voltage is delivered by rectifier 1 in normal operation. This operating condition is maintained for an adjustable time (e.g. 30 min to 3 h).

3a Supply to load 1: Rectifier 1 supplies load 1. The compensator is switched off some 5 min after return of the mains and bridged with contactor K2. Decoupling diode V6 prevents charging of the battery by rectifier 1.

3b Supply to load 2 and parallel mode with the battery: Rectifier 2 supplies load 2 and the battery.

4 Power failure time decision: If the duration of the power failure (2) has been shorter than 2 min and the parallel mode time is at an end, then a switch back to normal operation takes place (see 1). Battery discharge contactor K11 opens once the parallel mode time has expired, separating the battery from load 1. Rectifier 1 supplies load 1, as in mode 3a. Rectifier 2 is switched to trickle charging (2.23 V/cell) and thus feeds the battery. Load 2 is in parallel with the battery. In this way normal operation (see 1) is restored.

5 Charging: If the duration of the power failure (2) has been longer than 2 min and the parallel mode time is at an end, then there is a switch to charging of the battery (2.33 V/cell). Battery discharge contactor K11 opens and separates the battery from load 1. Rectifier 1 supplies load 1, as in mode 3a. Rectifier 2 takes over charging the battery with 2.33 V/cell. Load 2, connected in parallel with it, receives the same voltage.

After a time that can be specified (e.g. up to 24 h) rectifier 2 switches back to trickle charging (2.23 V/cell, see 1). Normal operating condition is thus restored.

3.6 Assignment of D.C. Supply Modes to Communications Systems

Table 3.2 reviews the modes mentioned in sections 3.3 to 3.5 in connection with examples of individual series of rectifiers, battery switching panels, control panels and compensators, as well as communications systems.

3.7 Further Modes

There are still some d.c. supply modes apart from those discussed in Sections 3.1 to 3.5. The most important of these will be presented in Sections 3.7.1 to 3.7.5.

Table 3.2 Assignment of d.c. supply modes to communications systems

Mode		Standby parallel mode						Changeover mode	
Equipment series/ equipment's rated current		GR 2/ 6, 12, 25 A	GR 2/ 40, 100 A	GR 3/ 20, 30, 100 A	GR 11/ 50, 100 A	GR 12/ 20, 30, 100, 200 A	GR 12/ 200, 500, 1000 A BF/ 1500, 2000 A	GR 10/[1]) 50, 100, 200 A BS/ 200, 400, 600, 1000, 2000 A	GR 10/ 500, 1000 A BF/2000, 3000 A SF
Communications systems		Power supply systems: rated current/lead battery							
EMD	48 V	6 to 25 A 25 cells			50 to 300 A 25 cells			600 to 2000 A 26 cells	
	60 V	6 to 25 A 31 cells			50 to 300 A 31 cells			200 to 2000 A 30 cells with AGE 31 cells without AGE	2000 to 10 000 A 30 cells with AGE 31 cells without AGE
ESK	48 V		40 to 300 A 25 cells		50 to 300 A 25 cells			600 to 2000 A 26 cells	2000 to 10 000 A 26 cells
EWSA	48 V				50 to 300 A 25 cells			600 to 2000 A 26 cells	2000 to 10 000 A 26 cells
	60 V							200 to 2000 A 30 cells with AGE[2])	2000 to 10 000 A 30 cells with AGE[2])
EWSD	48 V					20 to 600 A 25 cells	800 to 10 000 A 25 cells	600 to 2000 A 26 cells	2000 to 10 000 A 26 cells
EMS 600/12 000	48 V			20 to 300 A 24 cells					
KN system	48 V					20 to 600 A 25 cells			

3.7.1 Changeover mode with voltage gates

A variant of the changeover mode without interruption is the changeover mode with voltage gates (Fig. 3.13), also called the counter voltage technique. Instead of the tapping diode there are two groups of voltage gates V1 and V2 here for the uninterrupted changeover to battery discharging.

In normal operation with trickle charging rectifier 1 feeds the 60 V communications system with a voltage of 62 V. Rectifier 2 supplies the 30-cell battery with a trickle charge voltage of 67 V. Voltage gate V2 (charging level 2) in this operating condition is bridged by bridging contactor K2. Contact K1, on the other hand, is broken.

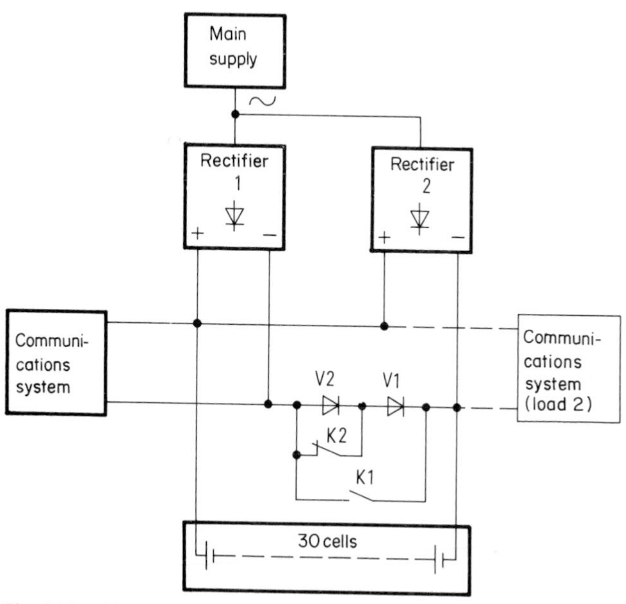

Fig. 3.13 Changeover mode with voltage gates (normal operation)

Notes to Table 3.2

GR Rectifier
BS Battery switching panel with control device
BF Battery switching panel
SF Control panel
AGE Compensator
EMD Noble-Metal Uniselector Motor Switching System
ESK Crosspoint Switching System
EWSA Analogue Electronic Switching System
EWSD Digital Electronic Switching System
EMS Electronic Modular System (PABX)
KN Communication Network (PABX)

[1]) GR 10/50, 100 A only for 60 V systems.
[2]) Combined standby parallel and changeover mode.

For this reason a voltage drop of some 5 V now occurs at voltage gates V1 (charging level 1). Only a small current is flowing. In this way the battery is prevented from feeding the communications system as long as rectifier 1 is delivering current.

In the event of a power failure contactor K1 makes contact and connects the battery to the communications system. The transition from mains to battery operation takes place without interruption, because the battery is feeding the communications system via voltage gate V1 during the switching time of contact K1.

In the operating condition charging rectifier 2 is changed over to a d.c. output voltage of 70 V. Now both voltage gates V2 and V1 are brought into action by the opened bridging contactors K2 and K1. A voltage drop of some 3 V also occurs at gate V2.

3.7.2 Parallel mode with 'floating' charge method

The parallel mode with 'floating' charge method is a variant of the parallel mode. In normal operation both rectifiers 1 and 2, both batteries 1 and 2 and the communications system are constantly in parallel (Fig. 3.14). A voltage of 2.05 to 2.1 V/cell is fed to the battery (some 63 V for 30 cells).

Sometimes no reducing diodes are required with this mode, as the operating voltage of 63 V lies within the permissible tolerance range of many communications systems. With this method 'equalizing charging' of, for example, the 2.4 V/cell must be done at regular intervals. For this rectifier 1 can be switched to the charging bar. This also applies to battery 2. Rectifier 1 is switched to a higher d.c. output voltage and battery 2 is thus charged. During this time rectifier 2 and battery 1 continue to work normally.

After the equalizing charging of battery 2 there is a switch back to normal operation.

3.7.3 Parallel mode with reduced number of battery cells and compensator

In this mode a 28-cell battery is used, for example, instead of one with 30 cells. The resultant trickle charging voltage lies within the tolerance range of most communications systems (Fig. 3.15).

In normal operation the rectifier supplies the communications system and the 28-cell battery with the trickle charging voltage. The compensator in this mode is bridged by contactor K.

In the event of a power failure the compensator is cut in by breaking contact K. The lower the battery voltage becomes, the more the (automatically controlled) additional voltage rises. Thus the voltage at the communications system remains constant even when the battery is discharging (e.g. with the 60 V system at about 61 V).

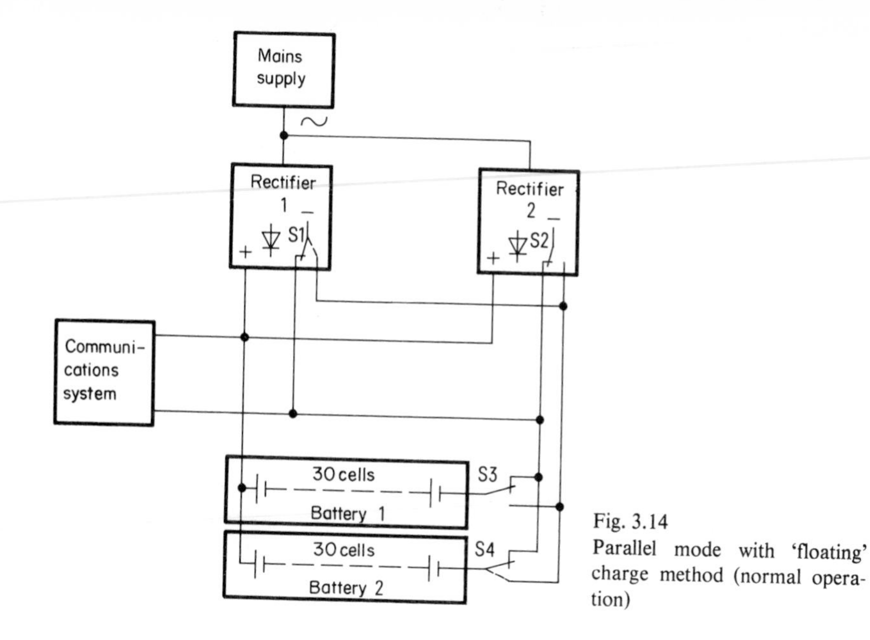

Fig. 3.14
Parallel mode with 'floating' charge method (normal operation)

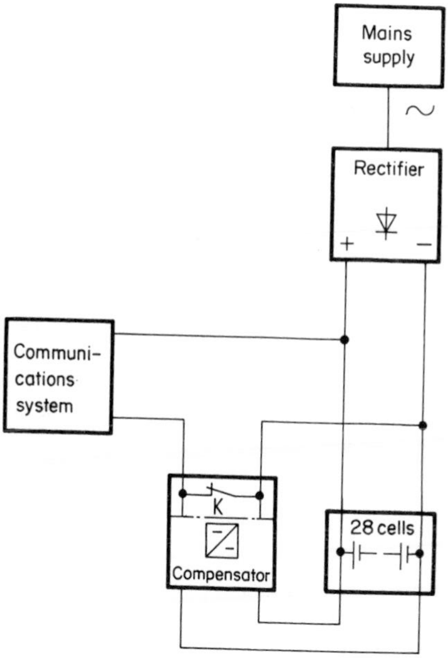

Mains supply

Rectifier

+ –

Communi-
cations
system

K

Compensator

28 cells

Fig. 3.15
Parallel mode with reduced number of battery cells and compensator (normal operation)

3.7.4 Parallel mode with compensator for additional and reverse voltages

There is another variant of the parallel mode with compensator using a 30-cell battery and designing the compensator for *additional and reverse voltages* (Fig. 3.16). The compensator is then operating *all the time.* It has to equalize voltage differences between communications system and the battery on trickle charging, charging and discharging.

In normal operation with trickle charging, the rectifier, as is well known, delivers a voltage of 67 V in the case of a 60 V system with a 30-cell battery.

For this reason the compensator must now generate a reverse voltage of 5 V so that the voltage at the communications system is not greater than 62 V.

In the event of a power failure and discharging of the battery the compensator delivers an automatically controlled additional voltage of up to 7 V.

On return of the mains and with charging of the battery the rectifier delivers a voltage of 70 V. In order that the voltage at the communications system does not become greater than 62 V, the compensator now generates a reverse voltage of 8 V.

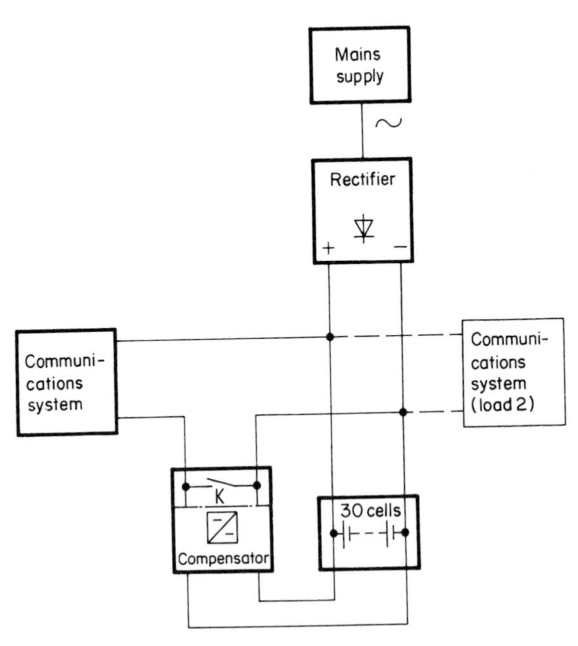

Fig. 3.16
Parallel mode with compensator for additional and reverse voltages (normal operation)

3.7.5 Parallel mode with end cells

The 'end-cell technique' (cubicle technique) is a variant of the parallel mode (Fig. 3.17). In normal operation the 'main rectifier' supplies the communications system and in parallel with it the 23 'main cells' of, for example, a 26-cell battery with trickle charging voltage via the made K1 contact (about 51.3 V). The 'end-cell rectifier' has only to supply the three end cells (in the case of trickle charging with some 6.7 V). In the event of a power failure contactor K1 opens and K2 closes, which thereby switches to battery discharging without interruption.

For charging the main battery cells the charging voltage must lie within the tolerance range of the communications system.

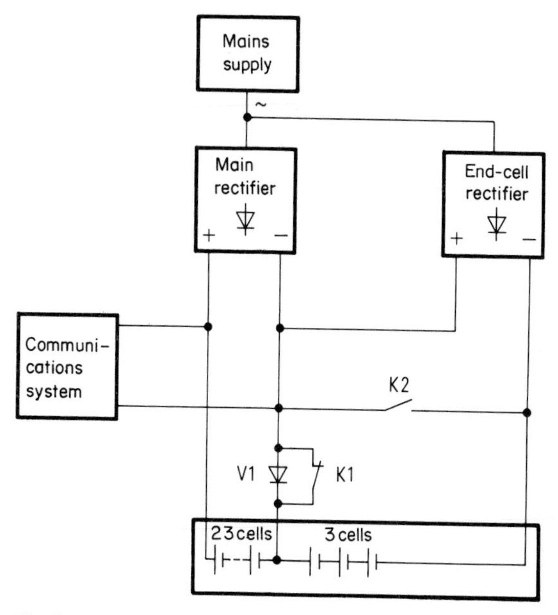

Fig. 3.17
Parallel mode with end cells
(normal operation)

4 Operating Modes of an Alternating Current Power Supply System

In addition to the direct current loads in communications systems there are also ones using alternating current such as computers and VDUs. The usual modes for their operation are shown in Table 4.1.

Table 4.1 Operating modes of an alternating current power supply system

Mode	System
Mains mode (without standby power supply)	
A.C. changeover mode with interruption > 1 s	Standby power supply systems Standby power supply available without limit on time with generator and internal combustion engine Starting mode
A.C. changeover mode with interruption <1 s	Rapid standby systems Standby power supply available without limit on time with generator and internal combustion engine Joint mode.
	Static standby power supply systems Standby power supply available with limit on time with inverter and battery [1]) Starting or joint mode
Uninterruptible a.c. mode with uninterruptible power supply systems	Immediate standby systems Standby power supply available without limit on time with generator, electric motor, flywheel and internal combustion engine Continuous mode
	Standby power supply systems with rotating converter Standby power supply available with limit on time with generator, d.c. motor and battery [1]) Continuous mode
	Static standby power supply systems Standby power supply available with limit on time with rectifier, inverter and battery [1]) Continuous mode

[1]) If necessary, additional standby power supply system. This then ensures an alternative power supply without any limit on time.

4.1 Mains Mode

In the mains mode the loads are supplied directly from the public mains (Fig. 4.1). There is no standby power supply.

4.2 A.C. Changeover Mode

In the event of a power failure the a.c. changeover mode provides a switch to an a.c. source standing by or running jointly (Fig. 4.2). Return of the mains results in switching back and renewed supply from the mains.

A distinction is made with standby power supply systems between those operating with an interruption >1 s and those with an interruption <1 s.

4.2.1 A.C. changeover mode with interruption >1 s

Standby power supply system

The standby power supply system is an alternative power supply system 'available without limit on time' with a generator and internal combustion engine (Figs 4.3 and 4.4).

In normal operation the loads are supplied directly from the mains. In the event of a

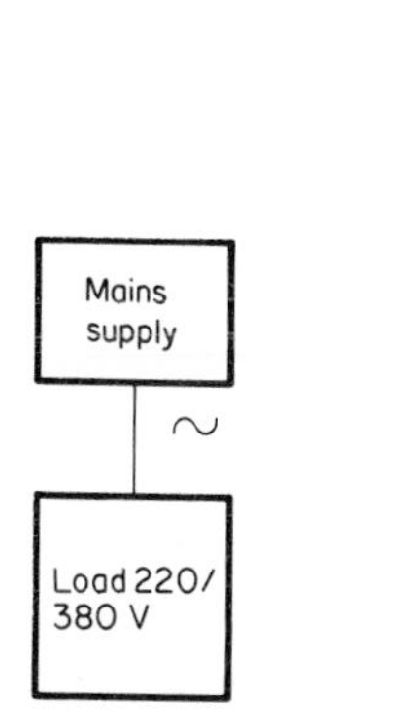

Fig. 4.1 Mains mode

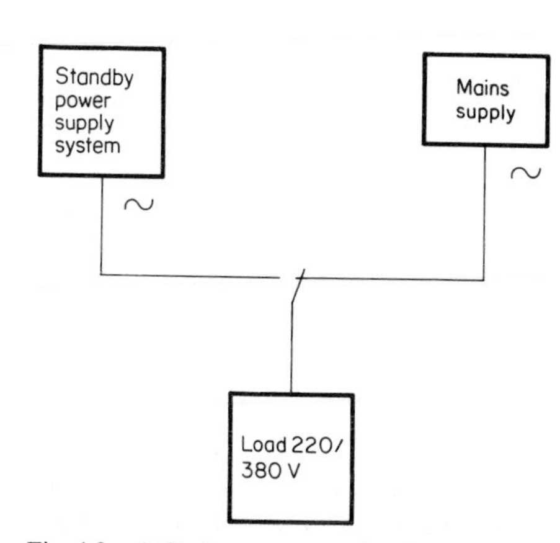

Fig. 4.2 A.C. changeover mode

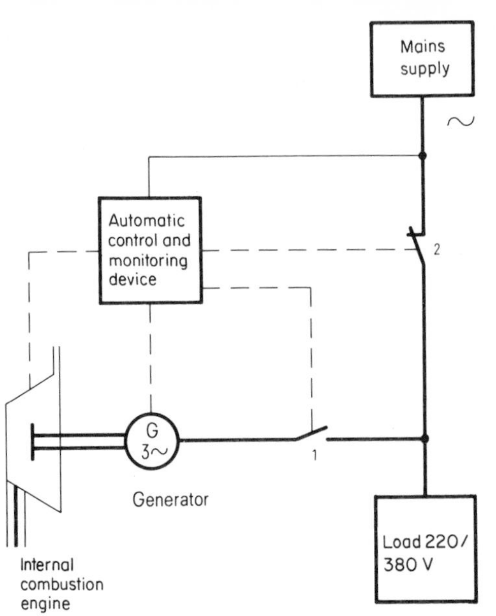

1 Generator contactor
2 Mains contactor

Fig. 4.3
Standby power supply system

Fig. 4.4
Standby power supply system
1250 kVA

power failure there is a changeover to the standby power supply system. For this it is necessary to start the internal combustion engine which is coupled with a generator. Once the rated speed has been reached the load is further supplied with alternating voltage via the generator. After return of the mains there is a switch back to mains operation.

4.2.2 A.C. changeover mode with interruption <1 s

With the aid of additional facilities operation can be maintained with a rapid changeover to the alternative a.c. supply 'without impairment of the load'.

Rapid standby system

An alternative power supply system with a generator and internal combustion engine 'available without limit on time' is known as a rapid standby system. It

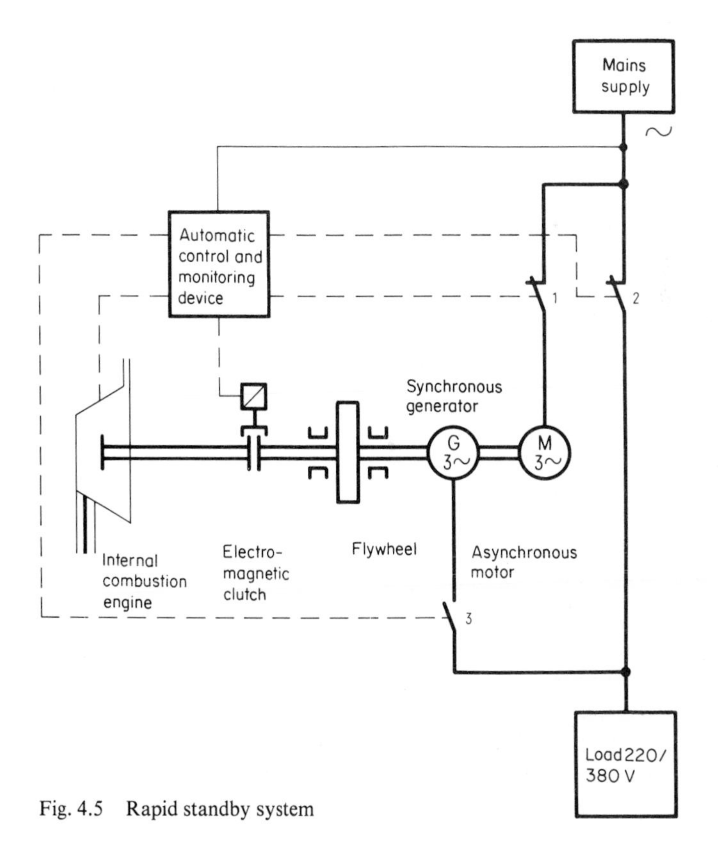

Fig. 4.5 Rapid standby system

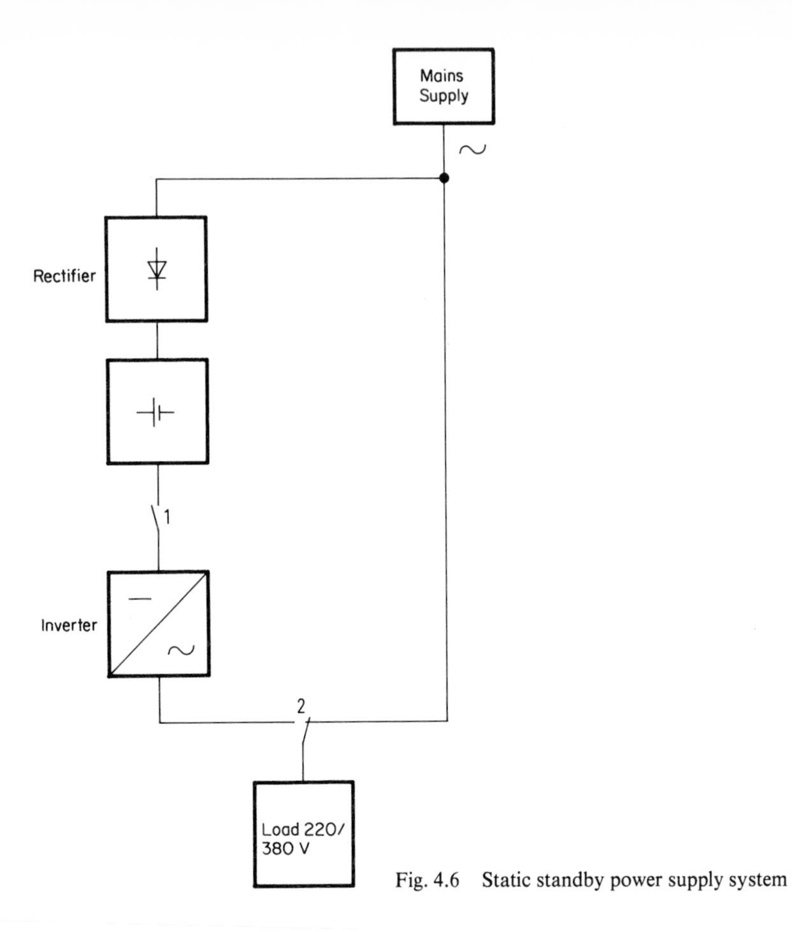

Fig. 4.6 Static standby power supply system

enables loads to be supplied which permit interruptions of max. 0.2 to 0.3 s in the event of a power failure (Fig. 4.5).

In normal operation the mains supplies the load and an asynchronous motor. The latter keeps the synchronous generator and flywheel at a speed just below the generator's rated speed. The flywheel can be separated from the internal combustion engine by means of an electromagnetic clutch.

In the event of a power failure a control system breaks contacts 1 and 2, makes contact 3 and starts the internal combustion engine. The flywheel is connected to the engine by means of the clutch. The energy stored in the flywheel is used to start the engine. In doing so the flywheel is also driving the synchronous generator. The engine now running at its rated speed together with the synchronous generator takes over and supplies the load. After return of the mains there is a switch back to normal operation.

Static standby power supply system

The static standby power supply system is considered as one with an inverter and battery that is 'available for a limited time'.

In normal operation the mains supplies the load directly and the battery is on trickle charging (Fig. 4.6).

In the event of a power failure the battery is switched to the inverter. This ensures a continued supply to the load. The supply is interrupted for a short time with any changeover operation.

4.3 Uninterruptible A.C. Mode

In the case of uninterruptible power supply systems (no-break a.c. power supply systems) the loads are supplied continuously, i.e. without interruption.

Regardless of their form such systems can be designed as 'single-block systems', 'single-block systems with passive redundancy' or 'multi-block parallel systems with passive and active redundancy (Fig. 4.7).

Single-block system

Figure 4.7(a) shows a single-block system with manual by-pass. The block shown there consists of:

▷ motor, generator, flywheel, internal combustion engine
  or
▷ rectifier, battery, d.c. motor, generator
  or
▷ rectifier, battery, inverter.

In normal operation the mains supplies the uninterruptible power supply block. This supplies the alternating voltage to the 'safe' bar and thus to the load. For maintenance work the whole uninterruptible power supply system can be made dead by means of the manual by-pass.

Single-block system with passive redundancy

In the event of trouble with the uninterruptible power supply block there is a switch back to the mains without interruption with the aid of the revert-to-mains unit (Fig. 4.7b). There is also a manual by-pass here.

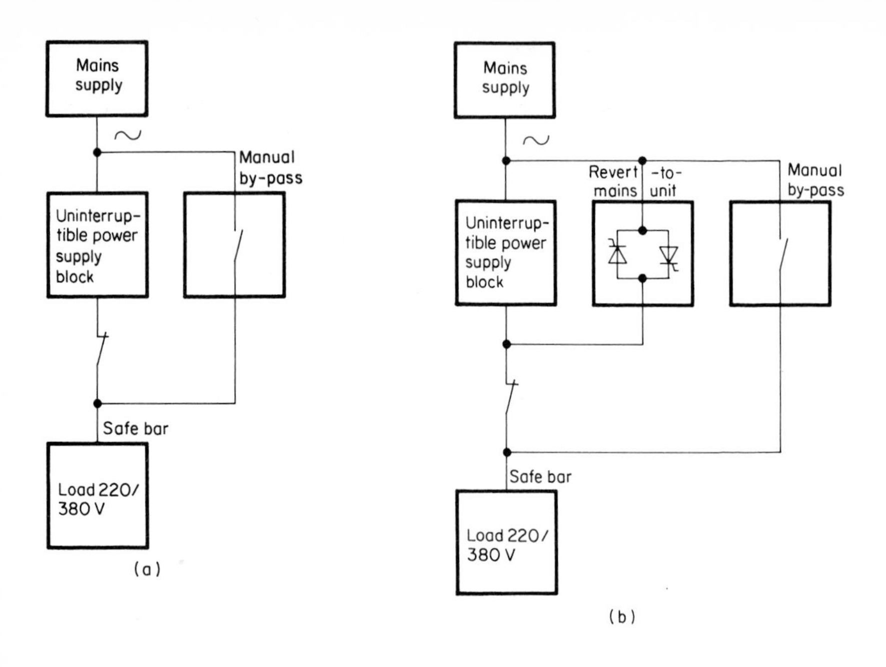

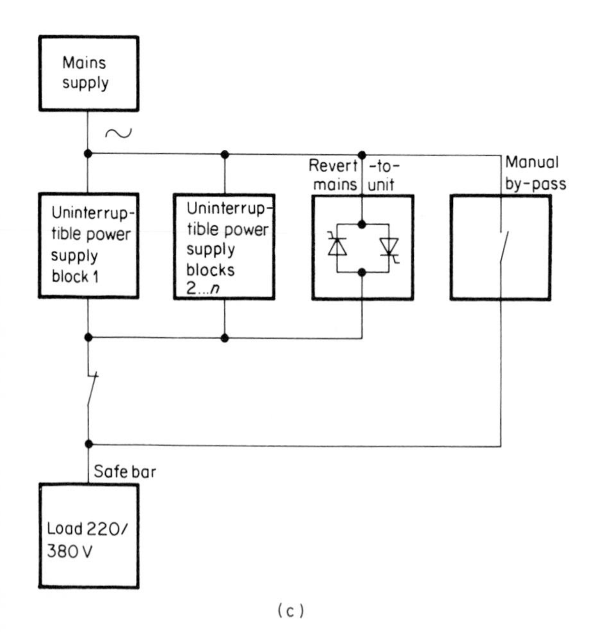

Fig. 4.7 Basic types of circuit of uninterruptible power supply systems

Multi-block parallel system with passive and active redundancy

This uninterruptible power supply system has active redundancy (Fig. 4.7c) in addition to passive redundancy (cf. Fig. 4.7b). There is active redundancy when the number of uninterruptible power supply blocks connected in parallel consists of at least one block more than is necessary for the load.

When there are a number of blocks in an uninterruptible power supply system, then, in the event of trouble with one of them, only the faulty block is separated from the 'safe' bar. In the example given there remains a further block which continues supplying the load.

If the output of the remaining block is not sufficient to supply the load or the second block is out of order, the revert-to-mains unit switches to the mains without interruption. There is also a manual by-pass with these systems.

4.3.1 Immediate standby system

The immediate standby system is often called a flywheel diesel converter system. It represents an alternative power supply without interruption and without any limit on time with a generator, electric motor, flywheel and internal combustion engine.

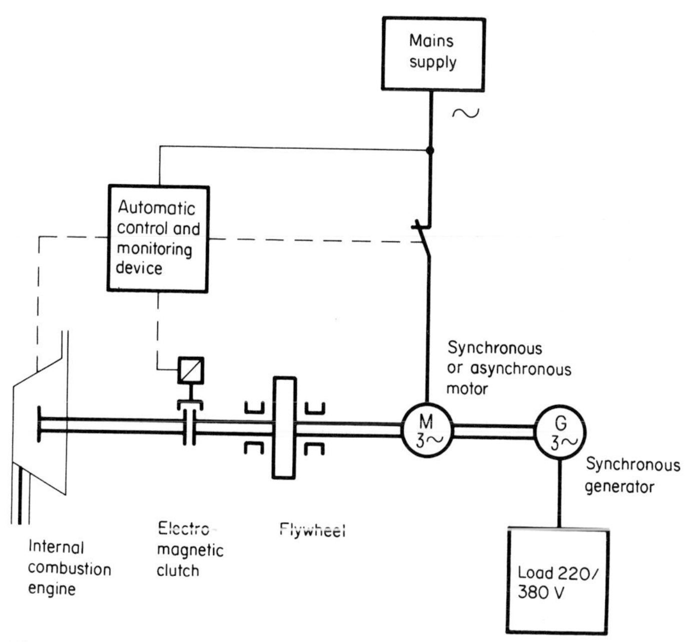

Fig. 4.8 Immediate standby system

Fig. 4.9 Immediate standby system 75 kVA

In normal operation the mains supplies a synchronous or asynchronous motor via the closed contact. The motor drives the synchronous generator and also keeps the flywheel at the rated speed (Fig. 4.8). The synchronous generator supplies alternating voltage to the load. The electromagnetic clutch is out and the flywheel is thus separated from the engine.

In the event of a power failure an automatic control system breaks the contact and lets in the clutch. The flywheel continues driving the synchronous or asynchronous motor and at the same time gets the internal combustion engine started. There now is a link between the engine and generator. The alternative power supply is ensured and is not limited in time. After return of the mains there is a switch again (without interruption) to mains operation.

Immediate standby systems (Fig. 4.9) are used, for example, for supplying power to radio link systems in telecommunications towers.

4.3.2 Uninterruptible standby power supply system with rotating converter

These systems are uninterruptible standby power supplies, available for a limited time with a generator, d.c. motor and battery.

Shown here as an example is a single-block uninterruptible power supply with rotating converter in the battery standby parallel mode, with passive redundancy through a revert-to-mains unit.

In normal operation the mains supplies a rectifier (Fig. 4.10) which drives a d.c. motor and at the same time feeds the battery. The d.c. motor is linked with a generator. The latter supplies the load via the cutoff unit.

In the event of a power failure the battery takes over the supply without interruption. The d.c. motor is then driven from power supplied by the battery instead of from the rectifier.

This is a standby power supply whose availability is limited in time by the capacity of the battery.

If trouble arises with the rectifier, battery, d.c. motor or generator, the contact of the cutoff unit is broken and there is a switch back without interruption to mains operation via the revert-to-mains unit. With this system there is also a manual by-pass as an additional safety measure. With this setup the whole uninterruptible power supply system can be isolated.

There can also be a standby power supply system in addition to the alternative power supply which is limited in time. In the event of a power failure there is a changeover to the standby power supply system. With this variant the battery

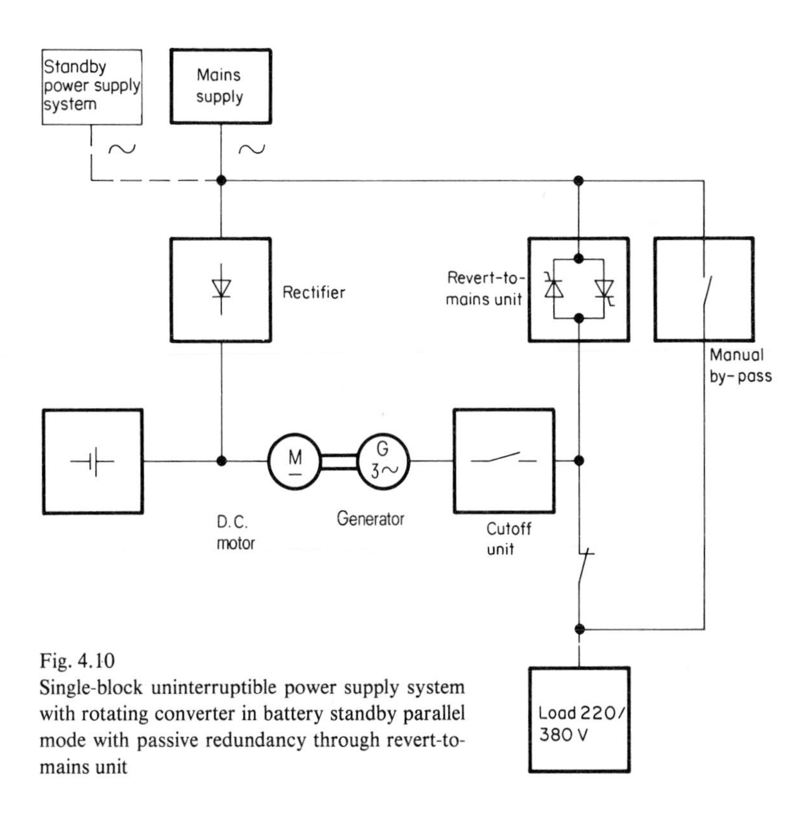

Fig. 4.10
Single-block uninterruptible power supply system with rotating converter in battery standby parallel mode with passive redundancy through revert-to-mains unit

capacity needs only to be designed for short bridging times. If there is also a standby power supply system, then there can of course be an alternative power supply without any limit on time.

4.3.3 Static uninterruptible standby power supply system

This system is considered as an uninterruptible standby power supply system, limited in time, with a rectifier, inverter and battery (Figs 4.11 and 4.12).

Static uninterruptible power supply systems are built up with batteries and, if needed, with standby power supply systems to further raise redundancy. Such systems are found, for example, in ground communications stations for telecommunications satellites, in the EDS communications system and in data processing systems.

Static uninterruptible power supply systems with inverters are characterized by a considerable improvement in efficiency compared with conventional systems. Because of a short settling time they also permit surges in load of up to 100% of the rated power, while maintaining all tolerances.

The reliability of such systems, which is substantial, can be further increased by raising the level of redundancy. It is well known that an improvement in active redundancy is achieved if the number of uninterruptible power supply blocks connected in parallel is increased by at least one block more than is normally necessary for the load.

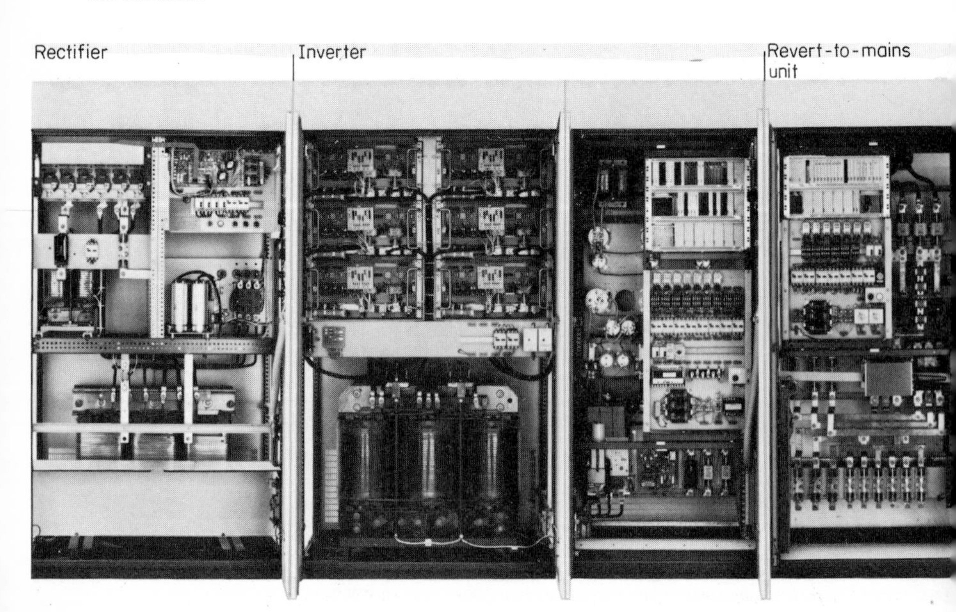

Fig. 4.11 Static uninterruptible power supply system 82 kVA

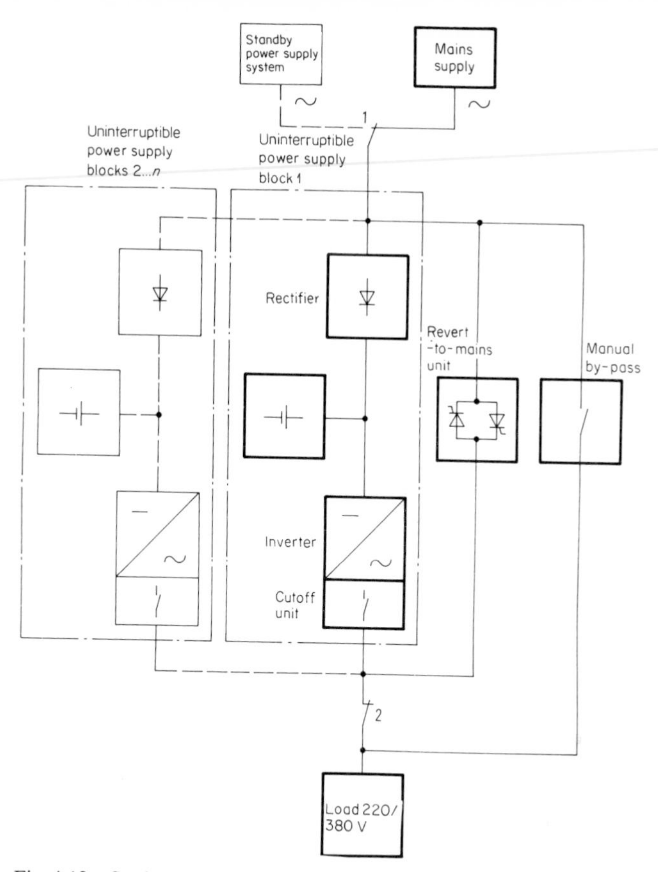

Fig. 4.12 Static uninterruptible power supply system in parallel mode

In normal operation the system works with all uninterruptible power supply blocks in the partial load range. It is only after the failure of a block that the remaining ones are fully loaded.

Passive redundancy is achieved with the aid of the revert-to-mains unit, which in the event of trouble directly switches the mains to the 'safe' bar and thereby to the loads.

Static uninterruptible power supply systems in parallel mode

In normal operation the rectifier is supplied from the mains. Within the uninterrupted power supply block this has to supply the inverter and the battery

(Fig. 4.12). The inverter is connected with the 'safe' bar (cf. Fig. 4.7) and thereby with the load.

As the battery is in parallel with the rectifier and inverter, it takes over and supplies the inverter without interruption in the event of a power failure or breakdown of the rectifier. The capacity of the batteries is usually selected for bridging times of 10 to 60 min.

Where necessary an additional standby power supply system provides an alternative power supply available without any limit on time. The revert-to-mains unit must not respond in this mode, as otherwise the standby power supply system would be switched directly to the 'safe' bar, which could result in frequency deviations outside the tolerance range for the load.

5 Public Mains – Conditions and Requirements

It is important to know the characteristics of the public power supply system (distribution network) when operating telecommunications power supply systems. This chapter explains, in a simplified form within the framework of this publication, the factors to be considered and the line-side conditions for which the equipment is designed.[1])

The factors considered include:

- ▷ type of voltage,
- ▷ tolerances of the alternating supply voltage and supply frequency,
- ▷ wave shape and distortion factor of the alternating supply voltage,
- ▷ power failures.

5.1 Type of Voltage

Rectifiers up to about 25 A are usually built for connecting to single-phase alternating voltage (220 V/50 Hz) and rectifiers for current strengths from 25 to 1000 A for connecting to three-phase alternating voltage (380 V/50 Hz). The equipment can be adapted to different alternating supply voltages or supply frequencies.

Figure 5.1 contains the designations of the conductors, voltage particulars and the designations of the connections at the rectifier. The protective conductor PE is *not* connected to the rectifier in the case of telecommunications power supply systems with function and protective earthing.

The voltages of outer conductors L1, L2 and L3 of the three-phase alternating voltage have a phase shift of 120° in relation to each other. The voltage between any two outer conductors is called *delta voltage* (380 V), while that between one outer conductor and the neutral conductor N is called *star voltage* (220 V).

The amplitude of an alternating voltage is indicated by its root-mean-square (r.m.s.) value. This is lower than the peak value by a factor of 1.4.

---

[1]) This whole section is based on:
DIN 57160/VDE 0160/November 1981
DIN 41750, Sheet 4, Supplement/October 1973
DIN 40110/October 1975.

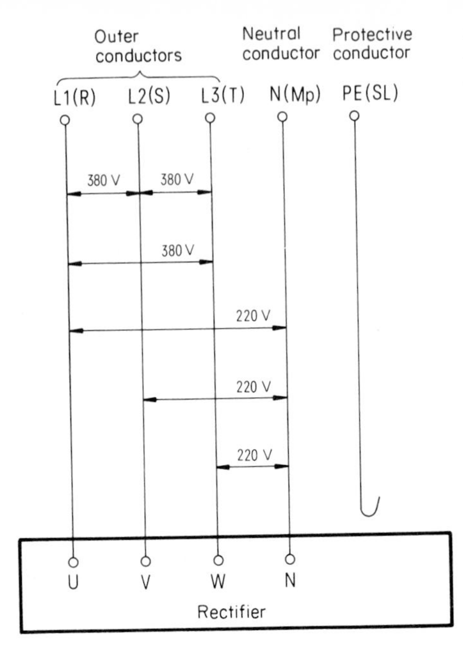

Fig. 5.1
Conductor designations and voltage particulars

5.2 Tolerances of the Alternating Supply Voltage and Supply Frequency

According to VDE, power converters are to remain serviceable when, in the long term, the alternating voltage fluctuates between 90 and 110% of the rated voltage (e.g. 220 or 380 V). Over and above this requirement rectifiers are normally designed for fluctuations in the supply ranging from −15 to +10%; i.e. as long as the supply voltage remains within this tolerance, the equipment's control system can keep the output constant at, for example, ±0.5% (static tolerance range).

Apart from the above voltage changes in the long term there also occur short-term, non-periodic undervoltages and overvoltages (voltage surges) in the mains supply. Atmospheric discharges can cause overvoltages up to 100 times the normal value for the mains voltage.

Such overvoltages are limited to permissible values in the low-voltage distribution system by means of surge dissipators. In addition, the varistors and TAZ diodes (cf. Section 7.3) built into the equipment provide further limiting. Overvoltages can also be due to switching operations in the mains network.

The interrelationship shown in Fig. 5.2 applies to non-periodic overvoltages as a departure from the rated peak value of the alternating supply voltage in the short term. Power converters must be designed so that their operation is ensured at over-voltages below curve 1. In the case of overvoltages within the range between curves

1 and 2, operation may be interrupted by the response of the protective devices, though there must be no damage to the converters.

Supply undervoltages (voltage drops) are also taken into account in addition to supply overvoltages. In the case of short-term undervoltages which can similarly occur, for example, with switching operations in the network, the converters must remain in operation for a time of max. 0.5 s, as long as the voltage does not fall by more than 15% of the rated supply voltage. In the case of larger or longer-lasting voltage drops the converters may switch off by means of a protective device.

The variance of the star voltage of the mains from the momentary value of the fundamental oscillation can be up to 20% of the peak value for the fundamental oscillation, as shown in Fig. 5.3. The time for permissible short-time intrusions in the alternating supply voltage is shown by width b; it depends on the permissible relative harmonic content (cf. Fig. 5.4). Such intrusions in the alternating supply voltage occur, for example, during commutating operations.

Frequency fluctuations in larger grid systems in West Germany are normally slight.

The control system of rectifiers, even in the case of frequency changes within the tolerance range of ±5% of the rated supply frequency f_N, keep the output voltage constant within the range of, for example, ±0.5%.

According to VDE, power converters must remain serviceable if the frequency of the public power supply system varies by up to ±1% from the rated value.

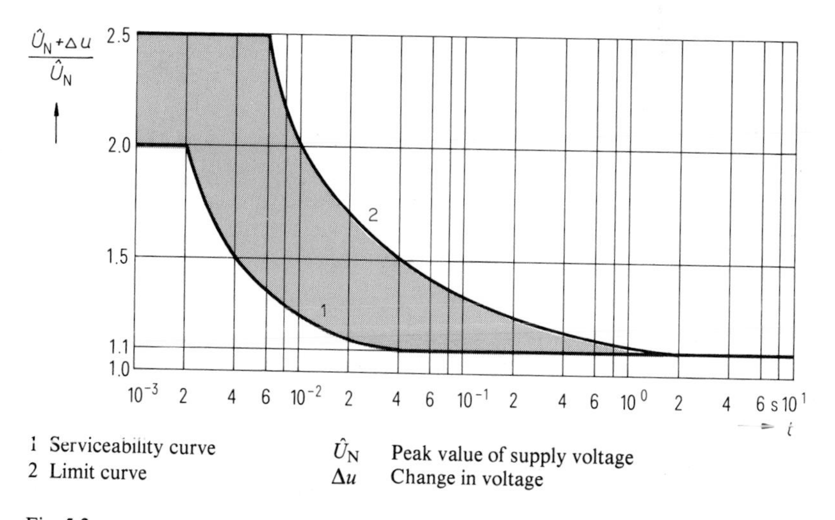

1 Serviceability curve
2 Limit curve

$\hat{U}_N$ Peak value of supply voltage
Δu Change in voltage

Fig. 5.2
Permissible non-periodic overvoltage of mains as function of time

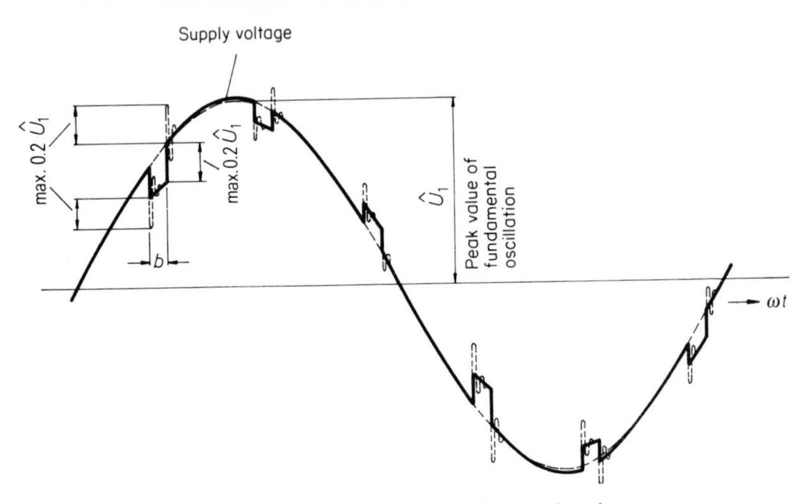

Fig. 5.3 Permissible short-time intrusions in alternating supply voltage

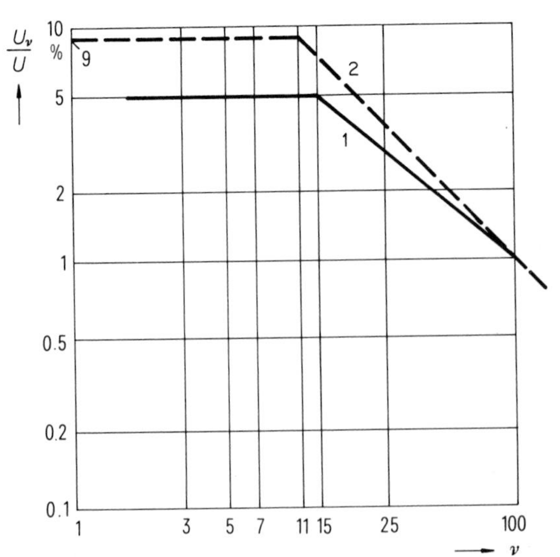

1 Continuous values
2 Short-time values
U_ν R.M.S. of the harmonic voltage of the νth order

Fig. 5.4
Limits for the harmonic oscillations in the alternating supply voltage

5.3 Wave Shape and Distortion Factor of the Alternating Supply Voltage

Loads connected to the mains, e.g. thyristor-controlled rectifiers with phase-angle control, generate harmonic oscillations and have a retroactive effect upon the mains, though harmonic waves also occur in the supply network itself.

For the perfect working of power converters the following applies to the mains:

- ▷ The relative harmonic content of the alternating supply voltage is max. 10%;
- ▷ for each harmonic the limit values as in Fig. 5.4, curve 1, are not exceeded as continuous values. No harmonic occurring, even for only a short time (time range of seconds), exceeds the limit values as in Fig. 5.4, curve 2;
- ▷ the largest periodic momentary value for the alternating supply voltage lies max. 20% above the respective peak value for the fundamental oscillation (cf. Fig. 5.3).

The *distortion factor* k represents a measure for the harmonic content:

$$k = \frac{\text{r.m.s. of the harmonics}}{\text{r.m.s. of the alternating quantity}}.$$

With ideal sinusoidal behaviour of the alternating supply voltage the distortion factor $k = 0$.

There also apply

$$k_{\mathrm{u}} = \frac{\sqrt{U_2^2 + U_3^2 + \cdots}}{U} = \frac{\sqrt{U^2 - U_1^2}}{U} = \sqrt{1 - g_{\mathrm{u}}^2};$$

$$k_{\mathrm{i}} = \frac{\sqrt{I_2^2 + I_3^2 + \cdots}}{I} = \frac{\sqrt{I^2 - I_1^2}}{I} = \sqrt{1 - g_{\mathrm{i}}^2},$$

where g is the fundamental oscillation content.

What harmonic currents occur depends on the rectifier's pulse number p (cf. Table 5.1). They have the frequencies νf with the ordinal number of the harmonic:

$$\nu = kp \pm 1 \quad (k = 1,2,3, \ldots),$$

where $\nu > 1$.

On the assumption that rectangular or stepped currents are taken from the mains (overlap angle $u = 0$ and ideal smoothing of the direct current), the amplitudes of the harmonics are in inverse proportion to their ordinal number.

The ideal harmonic current of harmonic number ν has, regardless of the degree of modulation, the r.m.s. value:

$$I_{\nu i} = \frac{1}{\nu} I_{L1i},$$

where

I_{L1i} is the basic oscillation of the ideal line-side conductor current.

The ideal basic oscillation content of the alternating current is:

$$g_{Ii} = \frac{I_{L1i}}{I_{Li}}.$$

The basic oscillation of the ideal line-side conductor current is, with a single-phase connection:

$$I_{L1i} = \frac{I_d U_{di}}{U_L}$$

and with a three-phase connection:

$$I_{L1i} = \frac{I_d U_{di}}{\sqrt{3} U_L}.$$

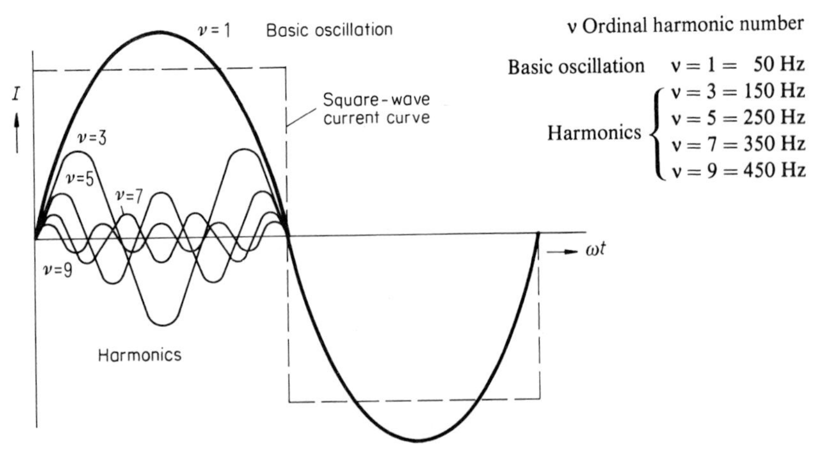

Fig. 5.5 Basic oscillation and harmonics of the mains supply

where

U_L is the conductor voltage
U_{di} is the ideal direct voltage
I_d is the direct current

With uncontrolled power converter circuits the phase position of the harmonic currents is predetermined, while with controlled ones the harmonic currents depend on the control angle α (cf. Fig. 5.7).

Every periodic oscillation can be shown as the sum of sinusoidal partial oscillations (Fig. 5.5).

The ideal square-wave current curve, which applies to a certain load in a sinusoidal network, contains a great number of components of different frequencies.

Table 5.1 contains all odd ordinal numbers ν from 1 to 25.

The supply frequency of, for example, 50 Hz is the basic oscillation (100%) and therefore contains the number ν = 1. The harmonics oscillate with integral multiples of the basic frequency (ν = 3 to ν = 25). The higher the frequency, the less is the amplitude of a harmonic wave, so that its disturbing influence also decreases.

Table 5.1
Ideal harmonic currents on the alternating current side of power converters

ν	ν*f*	Pulse number *p*		
		2	6	12
		$\frac{I_{\nu i}}{I_{L1i}}$ %		
1	50	100	100	100
3	150	33.33	—	—
5	250	20	20	—
7	350	14.29	14.29	—
9	450	11.11	—	—
11	550	9.09	9.09	9.09
13	650	7.69	7.69	7.69
15	750	6.67	—	—
17	850	5.88	5.88	—
19	950	5.26	5.26	—
21	1050	4.76	—	—
23	1150	4.35	4.35	4.35
25	1250	4	4	4

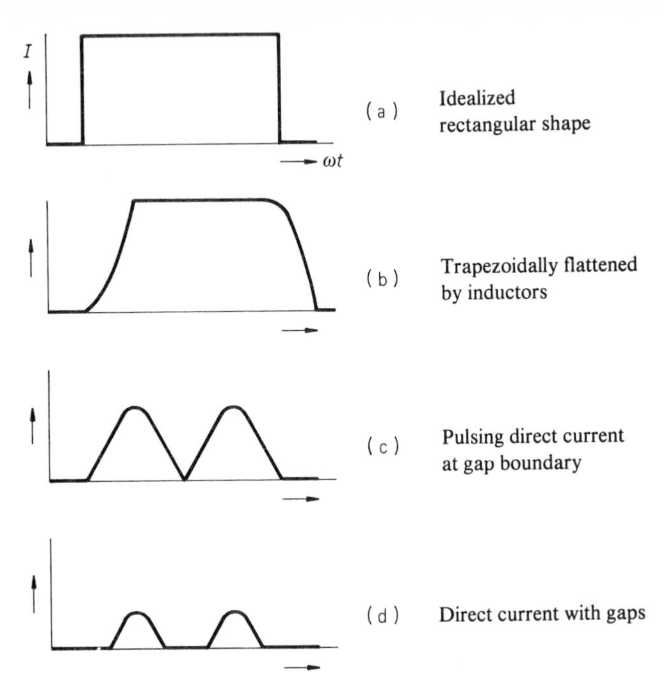

Fig. 5.6 Dependence of relative harmonic content on the shape of the current

The magnitude of the harmonic current is independent of the pulse number p and is the same in all circuits (if the frequency in question occurs at all in the higher-pulse circuits). For example the eleventh harmonic (ordinal number $\nu = 11$) with the frequency 550 Hz has with the three pulse numbers $p = 2$, 6 and 12 the same magnitude as the ideal harmonic current, namely 9.09%. It is therefore beneficial to raise the pulse number p, because the strongest current harmonics with low number and frequency do not then occur.

In thyristor-controlled rectifiers rated at up to about 25 A single-phase alternating current, double-pulse converter circuits are normally used in a semi-controlled single-phase bridge circuit (cf. Section 7.6.4). In thyristor-controlled rectifiers rated from about 25 A to 100 A the fully controlled (six-pulse) three-phase bridge circuit is normally used (cf. Section 7.6.4).

Six-pulse power converter circuits are also to be found in thyristor-controlled rectifiers from 200 to 1000 A (GR10), though in a fully controlled three-phase a.c. controller circuit (cf. Section 7.6.4). Through the additional use of a phase-shifting transformer in rectifiers from 500 and 1000 A a transition to a twelve-pulse arrangement can be made by operating the equipment in pairs (cf. Section 5.3.1).

Table 5.2
Harmonic currents with six-pulse fully controlled three-phase bridge circuit

ν	$\frac{I_\nu}{I_1}$	
	With square-wave shape of the current	With pulsing direct current at gap boundary
5	0.2	0.48
7	0.14	0.17
11	0.09	0.08
13	0.07	0.05

The ratio of the νth harmonic to the basic oscillation depends on the ordinal number, the control angle and the overlap. All line and thyristor side reactances cause a delay in current transfer between the commutating thyristors. The reactances flatten the current rise in the commutating circuit. The edges are then no longer ideally square but trapezoidal (Fig. 5.6). The amplitudes of the harmonics decrease with constant modulation as the influence of the reactances in the commutating circuit increases. The higher-order harmonics are attenuated more strongly by the influence of the overlap than are the lower-order ones. The larger the overlap angle u (i.e. the more inductances there are present in the circuit), the longer the trapezoidal semi-oscillations become. In this way the r.m.s. values of the actual harmonics are reduced. The overlap angle depends not only on the inductances in the circuit but also on the control angle α.

A continual change of the relative harmonic content can be assumed between the 'infinitely smoothed' direct current as in Fig. 5.6(a), and the 'pulsing direct current at the gap boundary', as in Fig. 5.6(c). The values for the fifth and sixth harmonics increase sharply as the gap boundary is approached, whilst the higher-order harmonics decrease.

The ratio of the harmonic current I_ν to the basic oscillation I_1 current with square-wave current and pulsing direct current with six-pulse fully controlled three-phase bridge circuit can be seen from Table 5.2.

The experimental values for ascertaining the maximum harmonic currents (on the supply side) to be expected from a controlled converter in fully controlled three-phase bridge circuit (six-pulse) are to be taken from Table 5.3.

The control angle $\alpha = 0$ (full modulation) is, as is well known, the same as that of an uncontrolled converter. Here with $u = u_0$ the commutation time is at its longest. The harmonics are thus particularly slight. There is the least modulation with the largest control angle α. This does not lead, however, to the maximum values for the harmonics. The maximum harmonic current occurs when there is roughly half-modulation.

Table 5.3
Maximum content of harmonic currents in r.m.s. current

ν	$\frac{I_{\nu\max}}{I_L}$ %
5	30
7	12
11	6
13	5

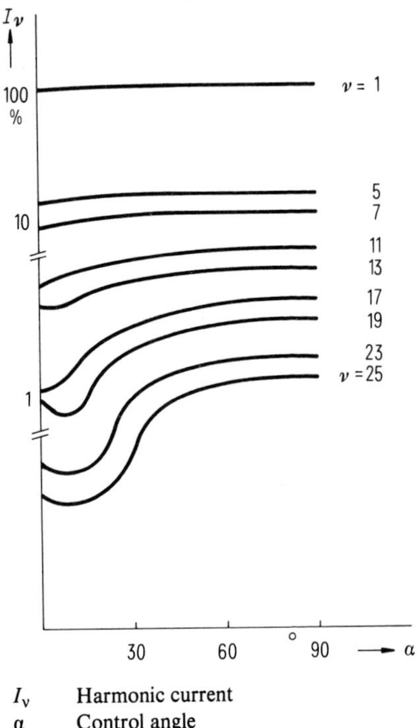

I_ν Harmonic current
α Control angle

Fig. 5.7
Dependence of the harmonic current on the modulation

Figure 5.7 shows the dependence of the harmonic current I_ν on the control angle α with the individual harmonic numbers ν (basic oscillation $\nu = 1 = 100\%$). In the example a six-pulse fully controlled three-phase bridge circuit has been selected.

5.3.1 Measures for reducing retroactive effects on the mains supply

There are various possibilities for reducing the occurrence of harmonics respectively suppressing their extension to the mains supply. The simplest course is to prevent harmonics occurring with the aid of a *circuit* (*with higher pulse number*). Another possibility is to reduce harmonics by means of *special control methods*. The spread of harmonics into the mains supply can also be reduced by means of *filter circuits*.

Suppression of current harmonics by raising the pulse number

A phase-shifting transformer is used in thyristor-controlled rectifiers (500 and 1000 A-GR10-equipment) for raising the pulse number and thereby reducing the retroactive effect upon the mains supply. Two rectifiers in each case form a 'pair'. In one the primary voltage is shifted by +15° and in the other by –15° (Fig. 5.8). The resultant retroactive effect upon the mains supply is similar to that normally used by a twelve-pulse circuit.

In Fig. 5.9(a) the phase-shifting transformer is connected with 15° lagging voltage and in Fig. 5.9(b) with 15° leading voltage. The phase-angle rotation is achieved by splitting windings with different numbers of turns between different sides of the

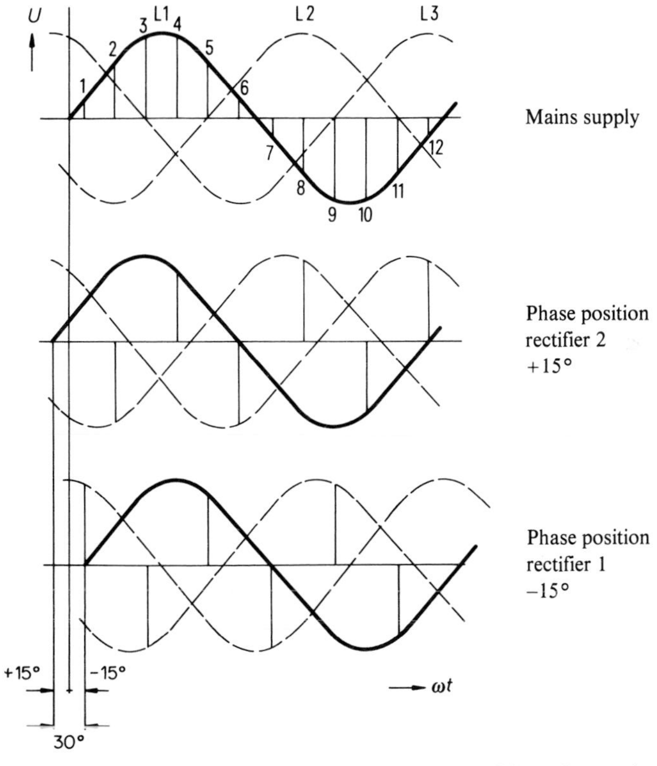

1 to 12 Firing points in relation to the phase position of the mains supply (12 pulses)
Example: Firing at 112°

Fig. 5.8 Phase shift due to phase-shifting transformer

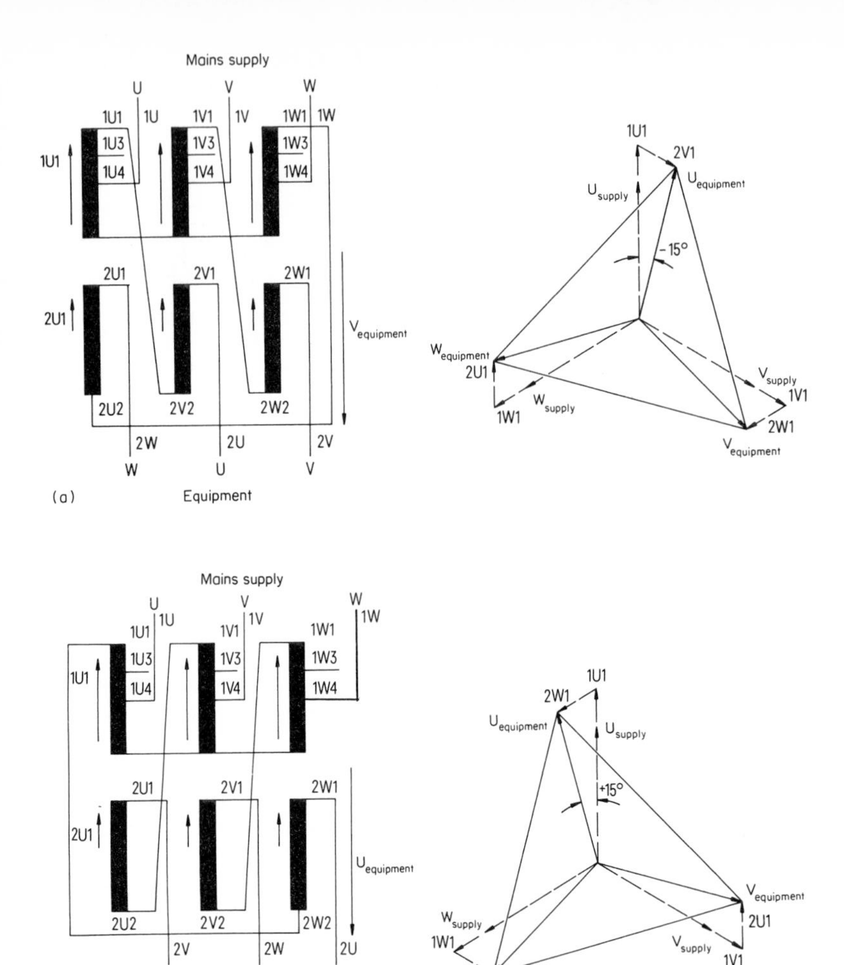

Fig. 5.9 Phase-shifting transformer

phase-shifting transformer. This transformer can also be used for adjusting the supply voltage in the event of values differing from 380 V.

With the aid of Fig. 5.10 it is possible to compare the stepped behaviour of the supply current of a six-pulse circuit (a) with that of a twelve-pulse one (b). A clear improvement can be seen with the twelve-pulse circuit compared with the six-pulse one, i.e. good approximation to the aimed at (ideal) sinusoidal shape of the current.

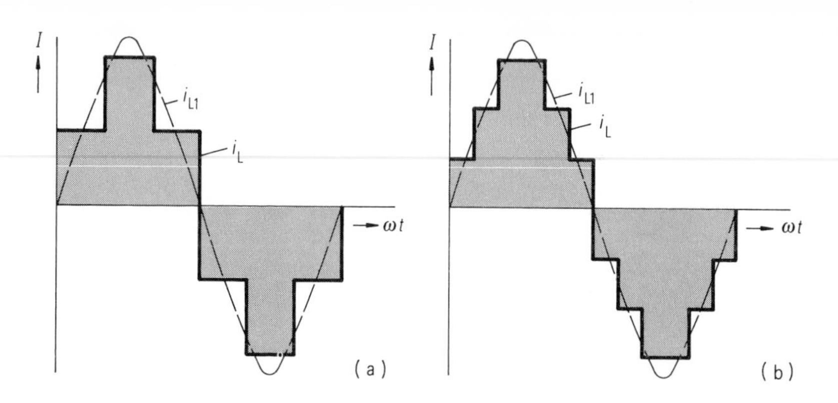

Fig. 5.10
Idealized, stepped behaviour of the mains supply current with power converters: a) six-pulse, b) twelve-pulse circuit

Suppression of current harmonics by oscillation package control

With oscillation package control (full oscillation control in cycle mode) complete full oscillations are blocked or passed. With this control system the thyristor is therefore not fired during certain oscillations. The more oscillations that are blocked, the smaller the output voltage becomes.

Suppression of current harmonics using filter circuits

Screening circuits, also called filter or absorption circuits, can be used to eliminate the harmonics on the low-voltage side caused by power converters. They are tuned to the frequency of the harmonics (Fig. 5.11).

In practice filters are normally used for the fifth, seventh, eleventh and thirteenth harmonics. A common absorption circuit, tuned to the twelfth harmonic, is assigned to the eleventh and thirteenth. In this way the harmonic currents can be reduced considerably.

If all ohmic resistances are disregarded and supply capacitances are excluded, the mains supply can be replaced by an inductance X_{LN}, which defines the short-circuiting power S_K of the supply at the converter connection point A. νX_{LN} is then valid for the effective reactance at the νth harmonic.

If filter circuits are connected to converter connection point A, thus in the immediate proximity of the converter, whose inductances and capacitances are selected so that a series resonance (and thereby a short-circuit) is produced for the harmonics to be suppressed, the currents of the frequencies concerned no longer

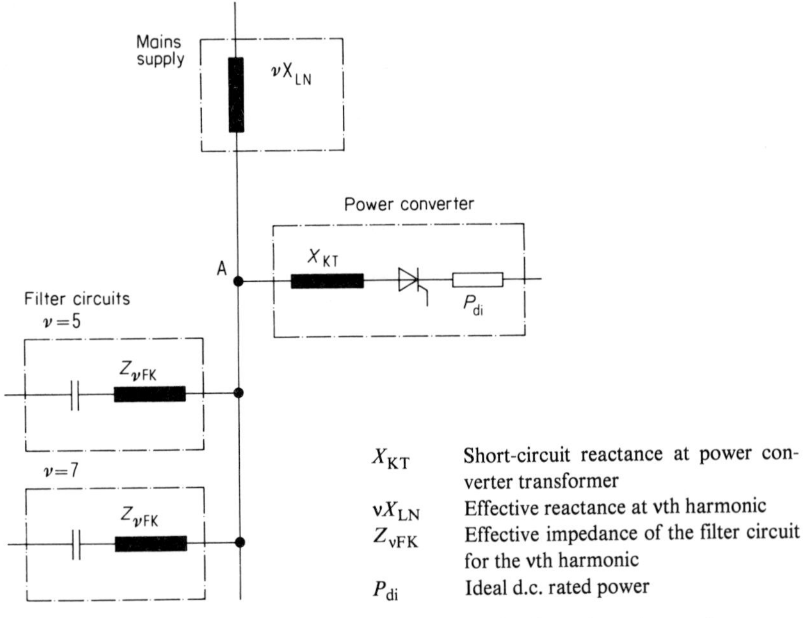

Fig. 5.11 Connection of the power converter and filter circuits to the mains supply

flow via the mains supply, but via the filter circuits. Below their tuning frequency the filter circuits work capacitively – for the basic oscillation too. They can therefore also be used for reactive power compensation of the system.

5.4 Power Failures

The reliability of public power supplies in West Germany is normally very high. However, there are 100 to 200 failures a year in the *short-time power failures* of less than 0.5 s. Such interruptions occur, for example, in the case of changeovers and automatic short-circuit reclosings.

Short-time failures in the millisecond range can possibly be bridged by the capacitors present in the rectifiers for filtering. In the event of power failures of longer duration, batteries, standby power supply systems, etc., are used.

In West Germany 2 to 4 *long-time mains failures* lasting a few minutes to several hours occur on average each year. Of these failures 97% are of the order of less than six hours and only 3% are longer.

There should also be included one interruption due to a fault with the service line and one due to maintenance work or faults in the mains high- and low-voltage switchgears, so that in all some six failures a year are to be expected with each system.

Table 5.4 shows the same time distribution for the duration of all interruptions as for mains supply interruptions. The table is based on 8000 telecommunications d.c. power supply systems.

Table 5.4
Frequency of interruption in the supply of power by telecommunications power supply systems

Duration of the interruption in power supply	Failures in %	Number of mains failures per year and power supply system	One fault per power supply system in years
Up to 2 min	40	2.4	0.4
2 to 30 min	25	1.5	0.7
30 to 60 min	14	0.85	1.2
1 to 2 h	9	0.55	1.8
2 to 4 h	8	0.5	2
More than 4 h	4	0.2	5
		Total 6.00	

6 Energy Stores

The battery (accumulator) as a reliable emergency power source for d.c. loads takes up electrical energy when charged and stores it as chemical energy. On discharging the chemical energy is reconverted into electrical energy.

There are *lead–acid batteries* (lead–acid accumulator) and *nickel–cadmium* (Ni–Cd) *ones*. They are made in different designs and with different properties. The fields of application are:

▷ fixed battery (stationary application),
▷ drive battery (mobile application),
▷ starter battery.

It is the fixed lead battery in particular that is important for a telecommunications power supply and is discussed further below (Fig. 6.1). Because of their different electrical characteristics, the individual designs of lead battery can be divided, in relation to different bridging times, as follows:

batteries for short-time loading (<1 h) and
batteries for long-time loading (capacitive loading, >1 h).

For short-time loading, for example, the designs GroE and OGi-Block are available, while the designs OPzS and OGi-Block are used for long-time loading.[1])

The consideration of other energy stores must be confined to an outline. Intensive work is taking place, for example, on the development of new energy stores and the further development of familiar ones. The *fuel cell* must be mentioned in this connection. It is not yet normally used these days in alternative power supply systems.

The idea of using fuel cells originates from the year 1930. Development has been pushed ahead in the last ten years primarily from the point of view of its use in electric vehicles, in space and for small power stations. Fuel cells work, for example, with hydrogen and oxygen. The fuels can be fed from pressure storages, the quantity of fuel stored being matched to the desired operating time. Fuel cells can also function as reversible units by the addition of electrolyzers, i.e. as when charging a battery water can be decomposed into hydrogen and oxygen, which are then available again when there is a need for energy.

There are also, for example, *sodium–sulphur* and *lithium–sulphur batteries*. The

---

[1]) OPzS – Fixed special design tubular plate battery.
OGi – Block fixed rod plate battery.
GroE – Fixed narrow design large surface area plate battery.

Fig. 6.1 Room with OPzS lead–acid battery (Photo VARTA AG)

energy densities currently achieved in prototypes are some four times higher than with conventional batteries. The disadvantage is that the electrochemical reaction (charging and discharging) will only take place at temperatures around 400 °C. The main aim of this development is to make available efficient drive batteries.

6.1 Lead–Acid Battery

Figure 6.2 illustrates the basic construction and the discharging and charging processes as well as the chemical equation for the lead battery. If electrodes consist-

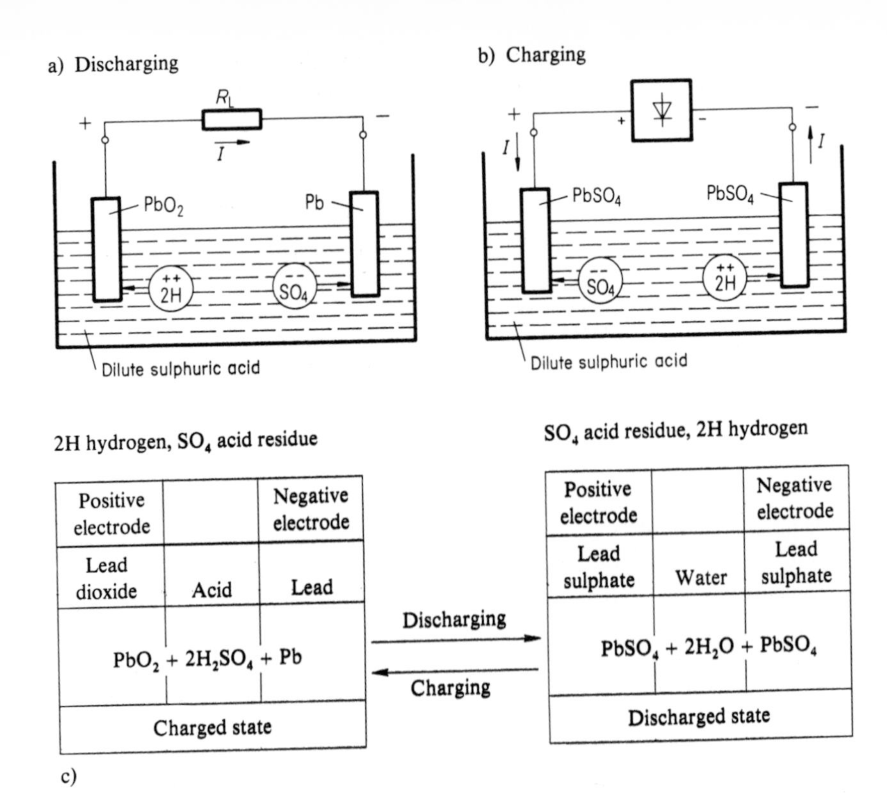

Fig. 6.2 Discharging and charging processes in lead–acid battery

ing of lead or lead compounds are immersed in a vessel filled with dilute sulphuric acid ($H_2SO_4 + H_2O$) as electrolyte, a secondary voltaic cell is produced. The active material lead dioxide (PbO_2) is produced electrochemically (formation) at the positive electrode. At the negative electrode the active material (lead oxide) is pressed into a grid of hard lead. Through forming, this paste is converted electrochemically into finely distributed spongy lead (Pb).

In a charged lead battery the active material at the positive electrode consists of lead dioxide (PbO_2) and at the negative electrode of lead (Pb).

6.1.1 Discharging

If the two electrodes are linked via a resistance (load, R_L), the current I flows (Fig. 6.2a). During this process the chemical conversion of the active materials of both plates is taking place.

Because of the electrochemical processes when discharging, both the lead dioxide of the positive plate as well as the lead of the negative plate are converted into lead sulphate ($PbSO_4$). Acid (H_2SO_4) is used up and water (H_2O) is formed, the concentration of the acid decreasing according to the energy drawn off. Discharging causes an initially slow, then quicker, drop in the voltage until a lower limit is reached (final discharge voltage), which is determined by the strength of the discharge current.

Conversion of the active, charged material into lead sulphate (discharging) involves a great increase in volume. During discharging the pores in the material become blocked and acid is prevented from reaching the inner particles of the material. This means an insufficient transfer of material (diffusion), a slowing down of the reaction and a drop in conductivity, which in the end leads to a fall in the discharge voltage until it drops below the set final discharge voltage.

6.1.2 Charging

The battery can be charged by connecting it to a d.c. source (e.g. rectifier), provided the voltage of the d.c. source is greater than that of the battery (Fig. 6.2b).

When charging, the active materials of the two electrodes and the sulphuric acid are restored to the original state existing before discharging. From a voltage of 2.4 V/cell the water briskly decomposes into hydrogen and oxygen (gassing). As this is harmful for the material of the plates in the long term, the strength of the charging current must not exceed certain values once the voltage at which gases form has been reached and must be reduced if necessary. The permissible strength of the current in the case of gases forming depends on the cell design and the method of charging. This voltage of 2.4 V/cell is not reached when using the battery in a telecommunications power supply, as the charging voltage is limited to 2.33 V/cell.

6.1.3 Open-circuit voltage

The open-circuit voltage of the off-load battery is also called the electromotive force (e.m.f.). This force depends primarily on the density of the acid (cf. Fig. 6.3). The higher the density, the higher the open-circuit voltage. It is sufficient in practice to know that the open-circuit voltage is approximately the same as the value for the rated acid density[1]) (charged state) plus 0.84.

6.1.4 Rated voltage

The rated voltage in the case of the lead–acid battery is 2 V/cell. It becomes established shortly after the start of discharging with the ten-hour current, when

[1]) In kilograms per litre at a temperature of 20 °C.

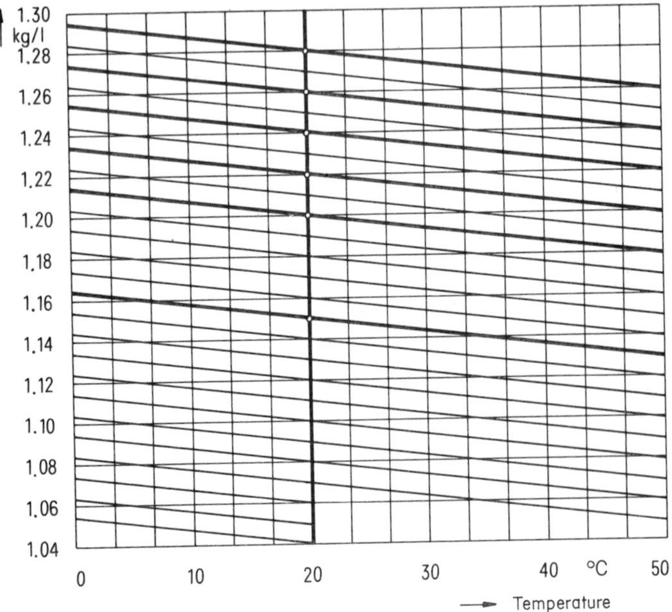

Fig. 6.3
Acid density as function of temperature

'inner' and 'outer' acids have balanced each other, i.e. once the voltage dip has been passed (cf. Fig. 6.4).

6.1.5 Discharge voltage

The voltage during discharging depends on the level of the discharge current and on the time. The higher the discharge current and the longer discharging lasts, the lower the voltage. The causes for this are the drop in acid density and with it the e.m.f., and also the additional voltage drop due to the anode plate resistance R_i.

For example, with OPzS batteries the acid density when discharging with the ten-hour current I_{10}[1]) starts at 1.24 kg/l and falls by 0.12 to 0.13 kg/l. When discharging with even larger currents the drop in acid density is less, corresponding to the lesser energy that can be taken off down to the permissible final discharge voltage.

[1]) I_{10} designates the current, in amperes, in relation to the rated capacity, with which a battery is discharged in 10 h down to a given final discharge voltage.

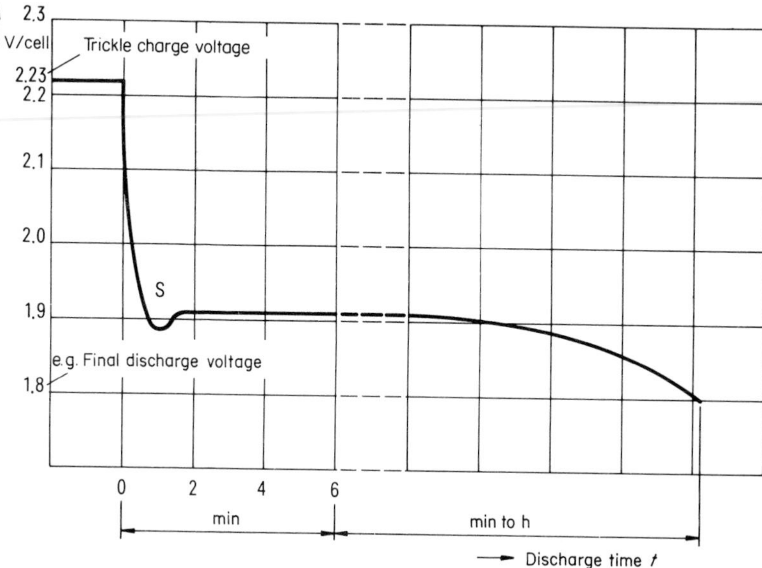

Fig. 6.4
Discharge voltage curve of a lead–acid battery

Taking into account the level of the discharge current, the state of discharge of the battery can thus be ascertained from the density of the acid. It should be noted that the density of the acid also depends on temperature.

A better measure for judging the state of discharge, however, is provided by the discharge voltage, taking into consideration the current strength in accordance with the discharge curves (cf. Fig. 6.5).

With rising temperatures the acid density drops by 0.0007 kg/l(per K) and rises correspondingly with falling temperatures. Thus with a change in temperature of 15 K one way or another there can be expected a change in density, in each case of around 0.01 kg/l in the opposite direction. Figure 6.3 facilitates ascertainment of the acid density values at different temperatures in relation to the rated temperature of 20 °C.

At the start of discharging the voltage passes through a minimum (voltage dip S, Fig. 6.4). One cause for the brief drop in voltage at the start of discharging is that the production of lead sulphate ions is temporarily delayed. Because of this unsteady behaviour the value always given as the initial discharge voltage is the one measured after removal of 10% of the capacity corresponding to the respective discharge current.

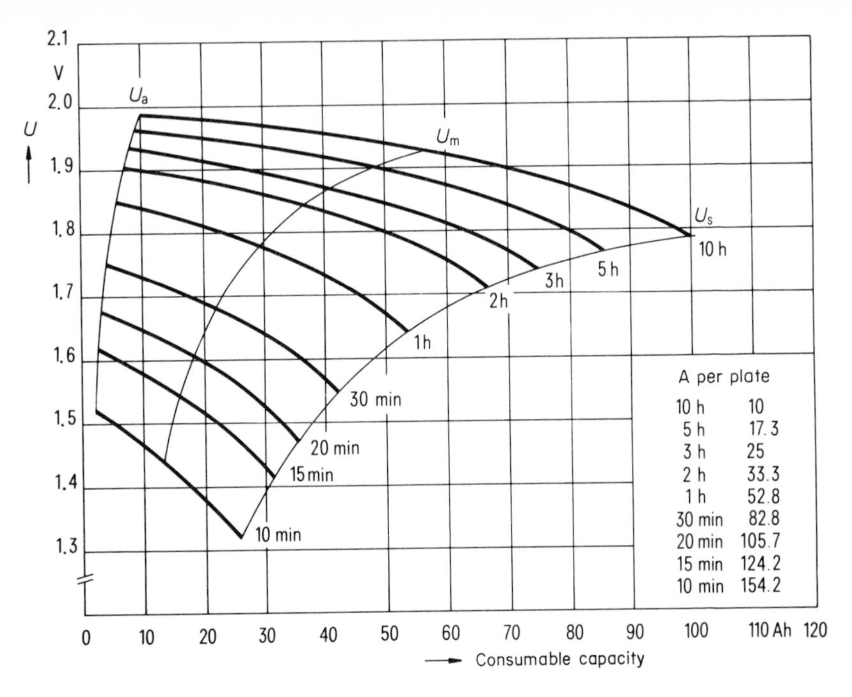

U_a Initial discharge voltage
U_m Mean discharge voltage
U_s Final discharge voltage

Fig. 6.5
Discharge voltage and consumable capacity of the OPzS 100 battery plate as a function of constant discharge current strengths

The behaviour of the voltage following the voltage dip is initially proportional to the decline in acid density. The voltage thus slowly drops in accordance with the characteristic curve of the battery in question.

Diffusion problems, depletion of active material and a decrease in conductivity also determine the further behaviour of the voltage with a continuing load. This is the reason why the voltage to the end of the discharge drops very rapidly the higher the discharge current, and less capacity can be taken from a cell until the final discharge voltage is reached. Thus the consumable capacity is reduced by some 50% if there is a one-hour discharge current instead of the ten-hour one.

Figure 6.5 shows as an example the discharge voltage of an OPzS battery and the consumable capacity. The rated capacity K_{10} here is 100 Ah. K_{10} accordingly indicates the amount of current in ampere-hours that, on discharging over a period of ten hours, can be consumed with the associated current I_{10}. Thus 10 A can be consumed for ten hours before the associated final discharge voltage U_s of 1.79 V/cell is reached.

When discharging with the current I_2 (= 33.3 A) a quantity of electricity of only some 50% of the rated capacity (K_{10}) could be consumed until there is a drop below the same voltage of 1.79 V/cell.

Final discharge voltage and system-conditioned final voltage

The final discharge voltage is the minimum allowable voltage level when discharging with the assigned current. If the voltage is allowed to fall below this level there is a danger that lead sulphate can no longer be reconverted, the efficiency of the battery thereby dropping sharply, respectively with frequent discharging below the final discharge voltage the structure of the active material in the plates is loosened or damaged by the changes in volume. Frequent falling below the final discharge voltage thus leads to a clear shortening of the battery's life.

The final discharge voltage (related to the battery) must not be confused with the bottom voltage limit for the communications system, below which it must not be dropped at the end of a bridging time. This voltage is called the system-conditioned final voltage $U_{s\,min}$.

6.1.6 Self-discharging

Due to the slow production of hydrogen at the negative electrode, lead–acid batteries are always subject to slight self-discharging. The usual phenomena of discharging can be seen here, namely sulphating of the plates and a fall in acid density. The loss in capacity occurring here can be up to 0.2% every 24 h at a temperature of around 20 °C. These losses must be compensated for by a continuous supply of energy.

This is why the battery is constantly operated on trickle charging in standby parallel and changeover modes.

Self-discharging is accelerated at higher temperatures.

6.1.7 Trickle charging voltage and charging voltage

During trickle charging the battery continuously receives a small current of 20 to 40 mA per 100 Ah of rated capacity. This charging current appears when a stable cell voltage (trickle charge voltage) of 2.23 V ± 1% is applied. This compensates for the losses in capacity described above due to self-discharging and keeps the battery in a fully charged state.

If the recommendations in the operating instructions are observed, lead–acid batteries can be run for their whole lifetime with a constant trickle charging voltage of 2.23 V/cell. Full charging is also possible with this voltage. To shorten the charging time it is also possible to apply accelerated charging, limited in time, at a higher voltage but below the voltage at which gas is formed (charging with 2.33 V/cell).

In the case of power supply systems with transistor and thyristor-controlled rectifiers, charging is normally '*time-dependent*' with 2.33 V/cell. The charging time is set with a timing relay (e.g. up to 24 h). At the end of charging there is a switch back to trickle charging.

The permissible charging current for the battery need not be limited even up to the point where gases develop reaching a voltage of 2.4 V/cell. However, if the temperature of the electrolyte rises above 55 °C, charging is to be interrupted.

A voltage higher than that at which gases form should not be applied without limiting the charging current, as there is a danger that the plates will swell. This can result in short-circuits in the cells.

An exception to the above maximum for the charging voltage is the voltage for initial charging (normally up to 2.7 V/cell with limiting of the charging current).

Table 6.1 shows for lead–acid battery types OPzS, OGi-Block and GroE the capacity available again after the respective times as a percentage, taking into account a charging factor of 1.2. The previously consumed capacity is replaced by charging in accordance with the *IU* characteristic curves. A trickle charging voltage of 2.23 V/cell or charging voltage of 2.33 V/cell is assumed.

In a power failure 50%, for example, of the battery's rated capacity (K_{10}) will be consumed. The full capacity is to be restored after return of the mains supply. If charging is done with 2.23 V/cell and a current of 0.5 I_{10}, then 90% of the capacity will have been put back after a charging time of 10 h and 97% after 20 h. This means for a 100 Ah battery (K_{10}) a current strength of 0.5 $I_{10} = 5$ A. Some days will pass, however, before the battery is again 100% charged (fully charged state). If, on the other hand, charging is with a voltage of 2.33 V/cell and a current of 0.5 I_{10}, then full charging (100%) is already reached after 20 h.

The behaviour of voltage and current when charging is finally determined by the size of the rectifiers and their characteristics. In standard DIN 41772 letter symbols are specified for the characteristics of rectifiers and thereby for the different methods of charging, and also for changeover and switching off processes. The meanings are:

U Constant-voltage characteristic
I Constant-current characteristic
W Sloping characteristic
O Automatic characteristic switching
a Automatic switching off after full charging

Table 6.1
Charging time with lead–acid batteries necessary after drawing off capacity in order to restore the desired level with *IU* charging (2.23 or 2.33 V/cell)

Consumption of K_{10} in %	Charging time t in h	2.23 V/cell				2.33 V/cell			
		$0.5\ I_{10}$	I_{10}	$1.5\ I_{10}$	$2\ I_{10}$	$0.5\ I_{10}$	I_{10}	$1.5\ I_{10}$	$2\ I_{10}$
25	3	87.5	94	95	96	87.5	95	97	97
	6	94	96	97	97	97	98	98	99
	10	96	97	97.5	97.5	100	100	100	100
	20	97	98	98	98	—	—	—	—
50	3	62.5	75	82	85	62.5	75	86	89
	6	75	89	90	92	75	92	94	95
	10	90	93	93	94	90	96	96	96
	20	97	95	95	95	100	100	100	100
75	3	37.5	50	62.5	68	37.5	50	62.5	75
	6	50	73	80	83	50	74	86	89
	10	66	86	88	89	66	89	93	94
	20	93	93	93	93	94	98	100	100
100	3	12.5	25	37.5	47	12.5	25	37.5	50
	6	25	50	65	70	25	50	76	80
	10	41.5	73	78	80	41.5	80	85	89
	20	80	91	91	91	82	95	100	100

K_{10} Capacity, ten hours. I_{10} Current, ten hours.

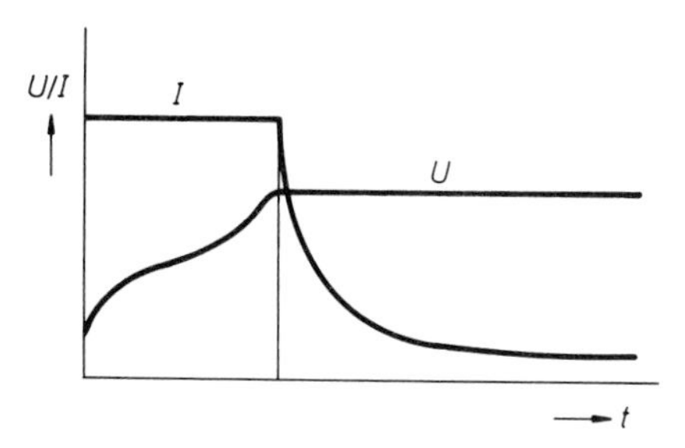

Fig. 6.6 Charging of lead–acid batteries for stationary systems in accordance with *IU* characteristic

Lead–acid batteries are charged in accordance with the *IU* characteristic by the rectifiers of the telecommunications power supply system (Fig. 6.6). This runs in two sections. First the charging current remains constant with a rising charging voltage until, depending on the characteristic chosen, the voltage of 2.23 V/cell or 2.33 V/cell is reached. From this figure on, the voltage is kept constant and consequently charging is with a current falling to lower values. When charging has ended there is, for example, an automatic switch back from 2.33 V/cell to 2.23 V/cell.

6.1.8 Initial charging voltage

It can be necessary on site to put lead–acid batteries on an initial charge (also called forming charge) with a charging voltage in the range from 2.6 to 2.8 V/cell. A voltage of 2.7 V/cell is usually chosen for initial charging.

'Special charging', e.g. in the case of heavy sulphating or after damage to the battery, is also carried out at up to this voltage.

The charging current specified for the battery, e.g. 0.5 I_{10}, is set by the rectifier and held constant during charging until the required cell voltage is reached. During charging the temperature of the acid must not exceed 55 °C.

Modern lead–acid batteries are supplied either filled and charged or unfilled and dry-charged. They therefore require only short initial charging. A large part of the capacity is immediately available after filling with acid.

With dry-charged cells it is quite possible when charging by the *IU* characteristic to carry out the initial charging at a voltage of 2.33 V/cell, unless the batteries have been stored for a lengthy time. In this case renewed forming charging must be carried out. It is assumed that at normal temperature and air humidity there is still a residual capacity of some 85% available after a storage time of 3 to 5 years.

There are instructions from the manufacturers of batteries on initial charging, which must be followed closely to achieve a long life for the battery.

6.1.9 Capacity

The capacity of a battery is a measure of its efficiency and size. It is measured in ampere-hours (Ah) and indicates the quantity of electricity a battery can deliver in a certain time (h) when discharging with a constant current (A) until a given voltage is reached (final discharge voltage).

Rated capacity means the rated value for the capacity with a specified discharge current strength plus the assigned final discharge voltage and temperature.

The greater the discharge current, the smaller the capacity and voltage. The rated capacity particular for all battery types is valid at an acid temperature of 20 °C.

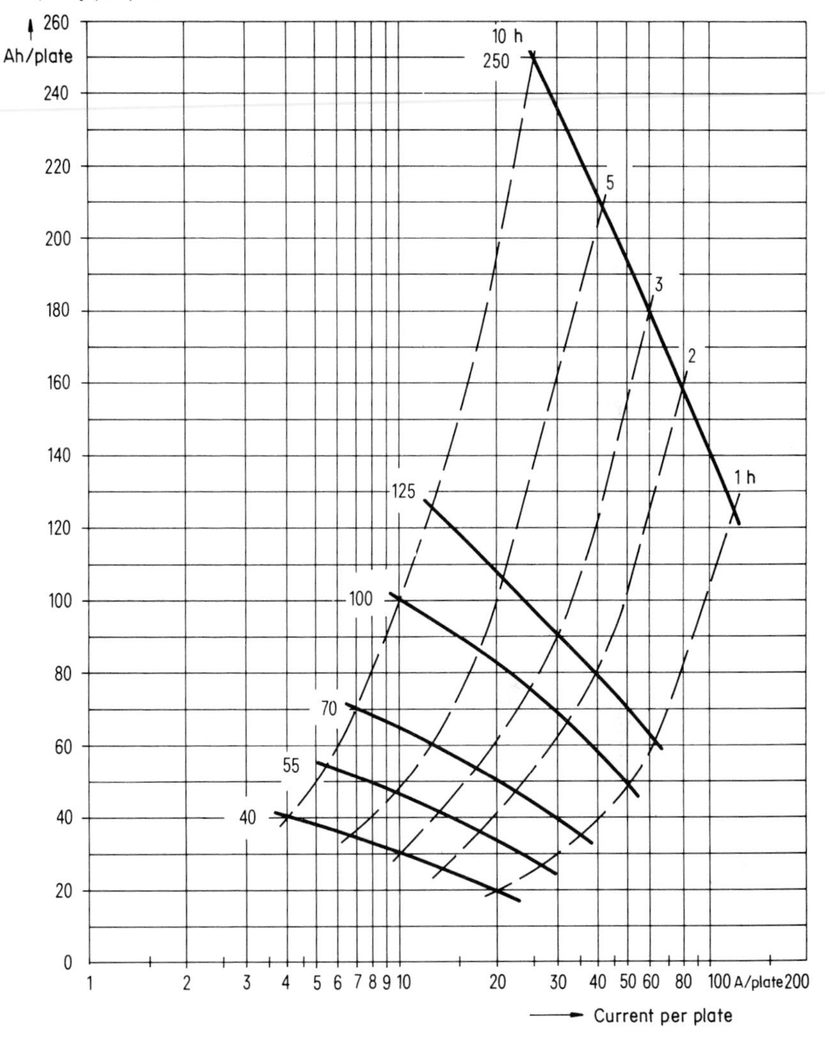

Fig. 6.7
Capacity curves for type OPzS lead–acid batteries (plates 40 to 250 Ah) at a temperature of 20 °C

At lower or higher temperatures the capacity is reduced or increased, within a limited temperature range by about 1% per 1 K. The full value for the rated capacity with new batteries is usually only achieved after three charge–discharge cycles.

The capacity curves in Fig. 6.7 show, for lead–acid batteries of the OPzS type, the respectively available capacities with different sizes of plate. The dependence of capacity on the discharge current strength can also be seen.

6.1.10 Efficiency and charging factor

A distinction is made between *ampere-hour efficiency* and *watt-hour efficiency*. These are shown as follows:

$$\text{Ampere-hour efficiency} = \frac{\text{consumed ampere-hours in Ah}}{\text{fed ampere-hours in Ah}}.$$

$$\text{Watt-hour efficiency} = \frac{\text{consumed watt-hours in Wh}}{\text{fed watt-hours in Wh}}.$$

Both figures depend on the cell construction, on the temperature of the acid and on the level of the charging and discharging currents. The ampere-hour efficiency is between 83 and 90%, the watt-hour efficiency at 67 and 75%.

By *charging factor* is understood the reciprocal of the ampere-hour efficiency. It is more customary in practice to write it as:

$$\text{Charging factor} = \frac{\text{fed ampere-hours in Ah}}{\text{consumed ampere-hours in Ah}}.$$

The charging factor is usually around 1.1 to 1.2.

7 Semiconductor Devices and Basic Circuits

P–N junction without external voltage

Figure 7.1 shows layer construction (a) and symbol (b) for the semiconductor diode.

The p–n junction, also called the depletion layer, occurs at the boundary of the p- and n-regions. If a p-region borders on an n-region then, due to the different concentrations of charge carriers, the free charge carriers want to balance out. Free electrons therefore diffuse from the n-region into the p-region; defect electrons (holes) pass in the other direction from the p-region to the n-region (Fig. 7.2).

The region between p- and n-layers is called the *space charged region* (Fig. 7.3). It is in this region that the exchange of charge carriers by preference takes place. Pair production and recombination processes already occur at room temperature without the application of external voltage.

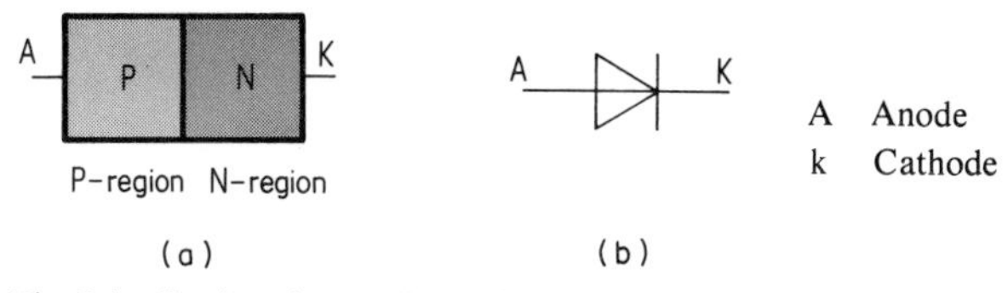

Fig. 7.1 Semiconductor diode: a) layer construction and b) symbol

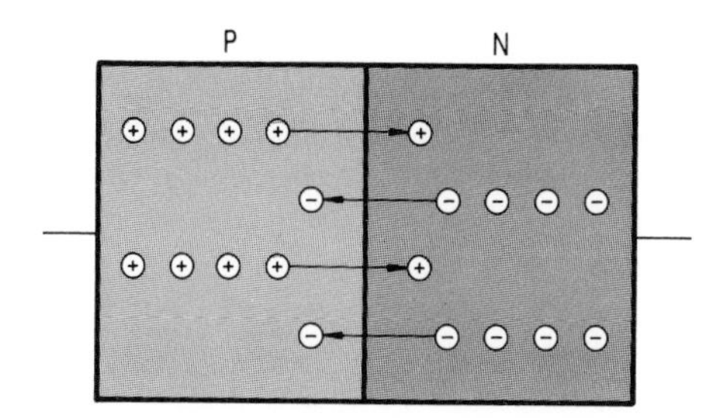

Fig. 7.2 P–N junction without external voltage

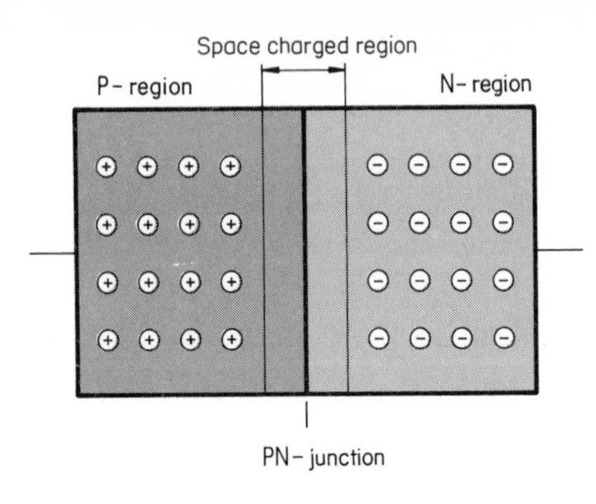

Fig. 7.3
P–N junction and space charged region without external voltage

Through a loss of holes the negatively charged acceptors remain uncompensated in the p-region. The p-region thus becomes *negatively* charged. After the release of free electrons positively charged donors remain uncompensated in the n-region. The n-region thus becomes *positively* charged. A voltage, the *diffusion voltage*, thereby forms between p- and n-regions. The diffusion voltage brings the balancing of free electrons and holes to a stop. A highly resistive layer depleted of free charge carriers is thus formed at the p–n junction.

P–N junction with external voltage

Diode in conducting state

Figure 7.4(a) shows the layer construction of a diode as regards area. The positive pole of a direct voltage source is connected with the anode and the negative pole with the cathode. The voltage of the direct voltage source must be higher than the diode's threshold voltage.[1])

By applying the positive battery voltage to the p-region the free charge carriers, in this case the holes, are driven across the p–n junction into the n-region (positive free charge carriers repel each other). Applying the negative battery voltage to the n-region similarly drives the free electrons by repulsion into the p-region. Thus from one moment to the other the p–n junction is flooded with free, moving charge carriers. As the depletion layer now disappears, the *diode now passes to the low-resistance (conducting) state.* Current now flows in the 'technical current direction' from the positive terminal of the battery across the anode/cathode space of the diode back to the battery's negative terminal.

[1]) The threshold voltage derives from the internal resistance of the diode, which results in an internal diffusion voltage.

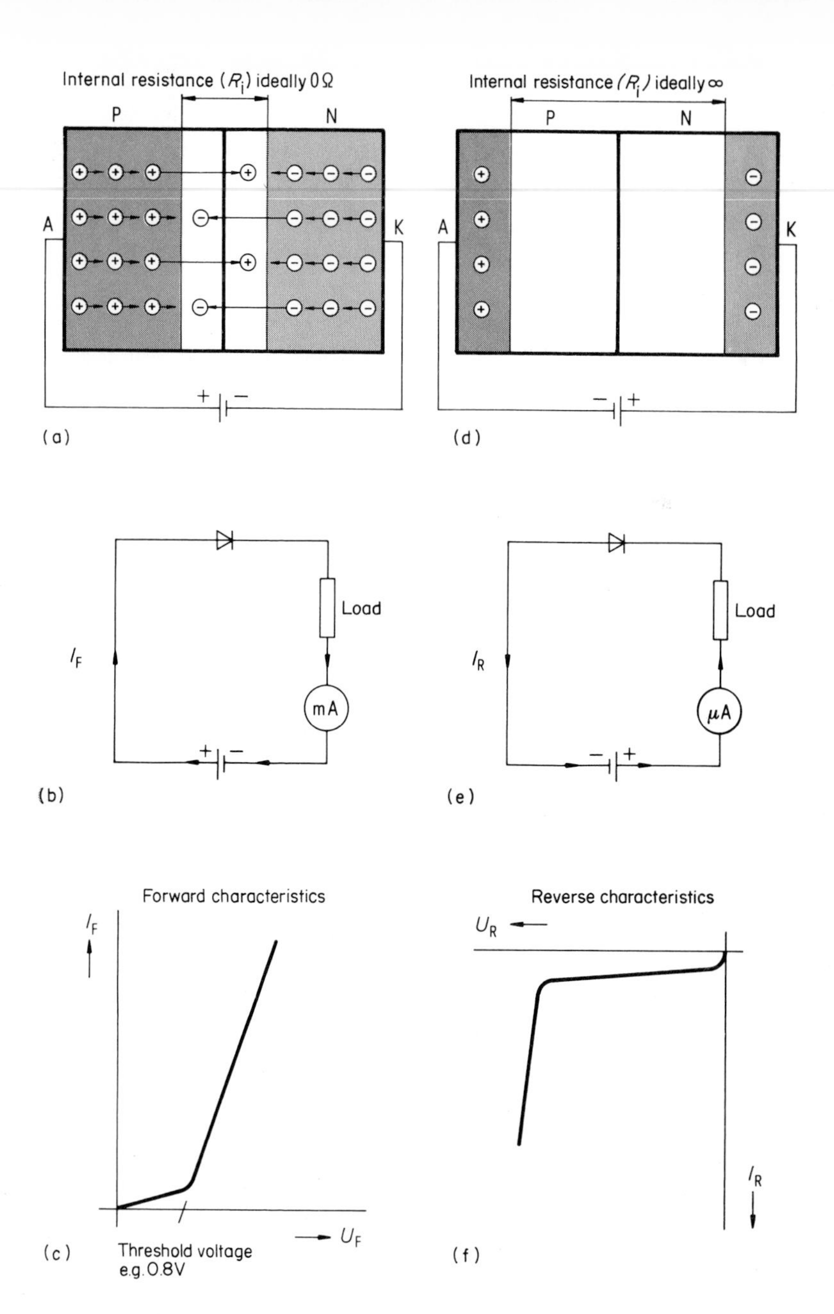

Fig. 7.4 P–N junction with external voltage (a and d) and semiconductor diode in conducting and off state

As Fig. 7.4(b) shows that the conducting state current is of the order of milliamperes.

Figure 7.4(c) shows the corresponding characteristic (forward characteristic) of the diode. After overcoming the threshold voltage the current begins to rise in a forward direction.

Diode in off state

In Fig. 7.4(d) the layer construction of the diode can again be seen as regards area. The poles of the direct voltage source have been reversed, the free charge carriers thereby being skimmed off. By applying the negative pole of the battery to the anode the positive free charge carriers are attracted to the p-region; connecting the positive battery pole to the cathode has the effect of attracting the negative free charge carriers. The space charge region thus expands. *The diode becomes highly resistive and passes to the off state.*

As can be seen from Fig. 7.4(e), there is now only a small reverse current I_R still flowing, and this is in the microampere range.

The reverse characteristic is shown in Fig. 7.4(f). The reverse current remains small despite the reverse voltage becoming ever greater. It is only when the 'breakdown voltage' is exceeded that the current rises very steeply. This can cause the device to lose its ability to function.

7.1 Selenium Diode

The selenium diode is a polycrystalline semiconductor. Because of its thermal inertia it can be greatly overloaded for a short time. It removes the problems from overcurrent protection. Nor is any special protection necessary against voltage peaks.

Selenium power rectifier plates are made for a rated reverse voltage of 25 V (PT) respectively 30 V (PU).

Figure 7.5 illustrates the typical *characteristic* for a selenium diode. If the voltage drop is less than 0.4 V in the forward direction, the diode remains blocked. No current flows. The threshold voltage is 0.4 to 0.6 V. The resistance value is not constant, becoming less as the voltage increases. Within a certain range the forward characteristic can be calculated with a constant voltage drop which is not dependent on the current flowing.

Figure 7.6 shows the typical, plate-shaped *construction* of a selenium diode. The rectifier plates can be combined into sets of plates (rectifier stacks). The plates are

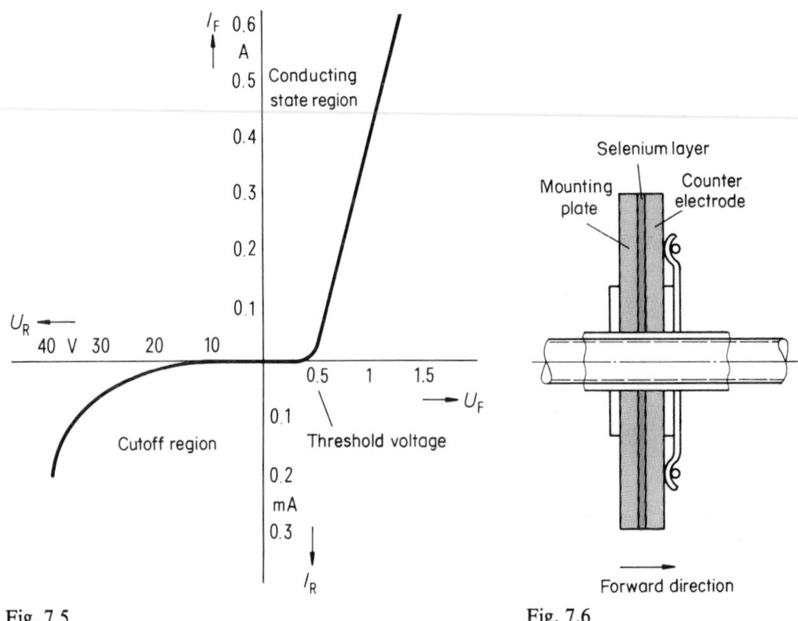

Fig. 7.5
Characteristic of a selenium diode

Fig. 7.6
Construction of a selenium diode

connected serially, in parallel or in parallel serially, depending on the level of the voltage, the level of the direct current and the type of circuit (Fig. 7.7).

In telecommunications power supply systems selenium rectifier stacks (e.g. as a tapping diode and decoupling diode) are usually used for *air self-cooling*. The tubular rivet type is also suitable for this type of cooling. There are also stacks for intensified cooling.

The highest, continuously permissible plate temperature is 85 °C.

7.2 Silicon Diode

The silicon diode is a single-crystal semiconductor.

As can be seen from the *characteristic curve* (Fig. 7.8), the threshold voltage is some 0.8 V. Below this there is virtually no forward current flowing. The forward voltage is 1 to 1.85 V.

The forward current is very small at low voltages and increases suddenly from a certain voltage.

Fig. 7.7 Selenium rectifier stack for air self-cooling

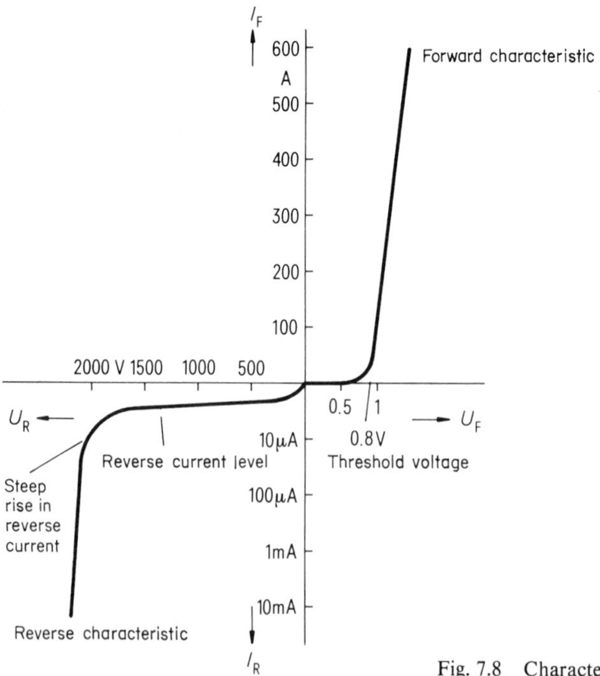

Fig. 7.8 Characteristic of a silicon diode

Fig. 7.9 Rectifier stacks with silicon insert diodes

The reverse current remains very small up to a very high voltage. Only when the breakdown voltage has been exceeded is there a steep rise in the reverse current.

There are the following *versions*:

▷ small diodes,
▷ insert diodes,
▷ screw diodes,
▷ flat-bottom diodes,
▷ disk diodes.

Figure 7.9 shows a selection of rectifier stacks with silicon insert diodes as used in thyristor-controlled rectifiers, e.g. as reducing diodes (counter cells).

Figure 7.10 shows silicon power diodes in screw, flat-bottom and disk design. Screw and disk diodes are primarily used as rectifiers in telecommunications power supply equipment.

Smaller diodes conduct the generated heat losses directly to the surrounding air. Larger designs require heat sinks. The permissible current loadability of the silicon diode can be increased up to threefold by switching from this *air self-cooling* (natural convection) to forced air cooling (forced convection). A further increase is possible with liquid cooling.

Fig. 7.10 Silicon power diodes

Normally silicon power diodes with air self-cooling are used in a telecommunications power supply.

Silicon power diodes are produced to the current state of the art for continuous limit currents up to 4200 A, for periodic peak reverse voltages of up to 4000 V and for maximum, continuously permissible depletion layer temperatures of 180 °C.

In contrast with the selenium diode, measures have to be taken against overvoltages (by means of commutation processes) in the case of the silicon diode, and similarly with the thyristor (cf. Section 7.6). The *carrier storage effect wiring* is used for this (Fig. 7.11).

As mentioned at the beginning, the p–n junction is flooded with free charge carriers when the diode is switched in the forward direction. In the case of reverse poling

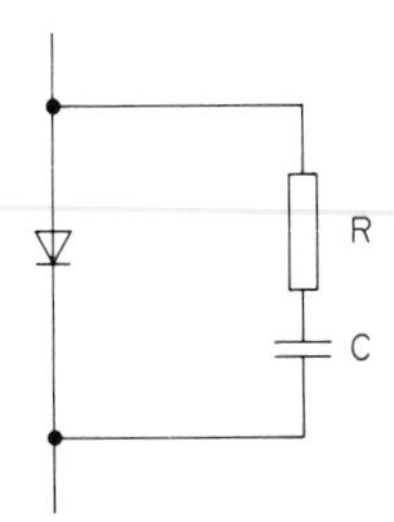

Fig. 7.11 Carrier storage effect wiring

there is a back current that is considerably higher than the normal reverse current. This phenomenon is known as the carrier storage effect. When the free charge carriers have disappeared, the back current suddenly collapses and voltage peaks arise through existing inductances.

7.2.1 Rectifying

One-way circuit

With the one-way circuit (Fig. 7.12) only the positive half-waves of an alternating current are passed. This circuit is suitable for *smaller* outputs. *More expensive* filtering is necessary to suppress interference voltages.

Centre circuit

Two transformer windings are to be connected in series for the centre circuit (two-way circuit, Fig. 7.13). Here both half-waves of the alternating voltage are used. As always, only one half of the transformer's secondary winding is carrying current, the transformer not being operated to its full extent. The centre circuit is therefore suitable only for *smaller* to *medium-sized* powers and *smaller* voltages. Less expensive filtering is necessary with the centre circuit than with the one-way circuit.

Bridge circuit

With the bridge circuit (Fig. 7.14) there is current flowing through two diodes simultaneously at each half-wave, while the two other diodes are polarized in the reverse direction. Here both half-waves are used, as with the centre circuit.

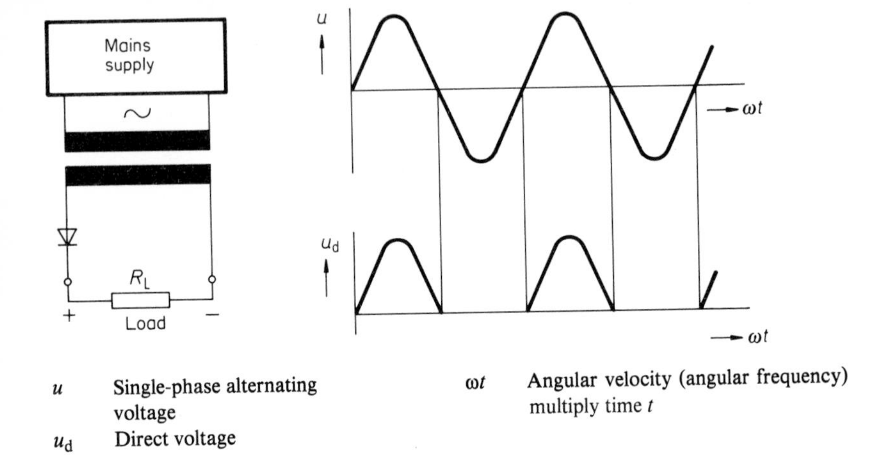

u Single-phase alternating voltage

u_d Direct voltage

ωt Angular velocity (angular frequency) multiply time t

Fig. 7.12 One-way circuit

The bridge circuit is suitable for *medium-sized* powers and *medium* voltages. It is similar to the centre circuit as regards expenditure on filtering.

Three-phase bridge circuit

The three-phase bridge circuit (Fig. 7.15) uses all three alternating voltages shifted by 120°. It is suitable for *large* voltages and *high* powers and requires a minimum of expenditure on filtering.

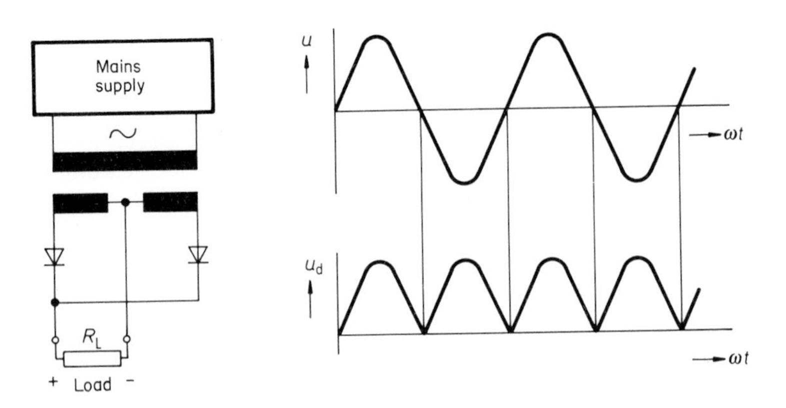

Fig. 7.13 Centre circuit

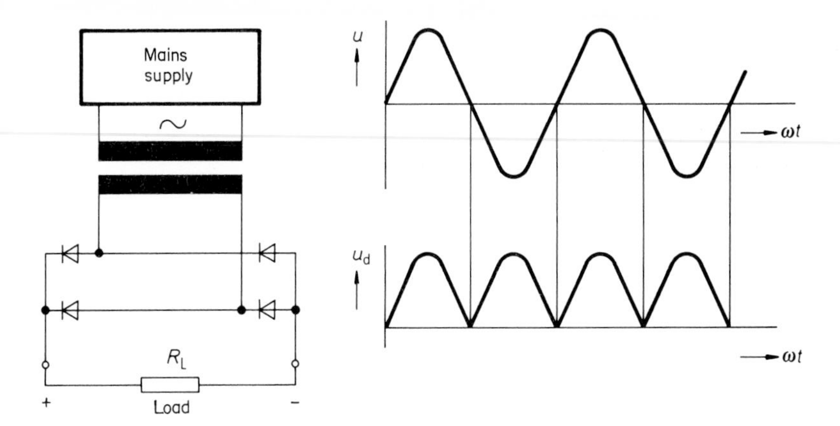

Fig. 7.14 Bridge circuit

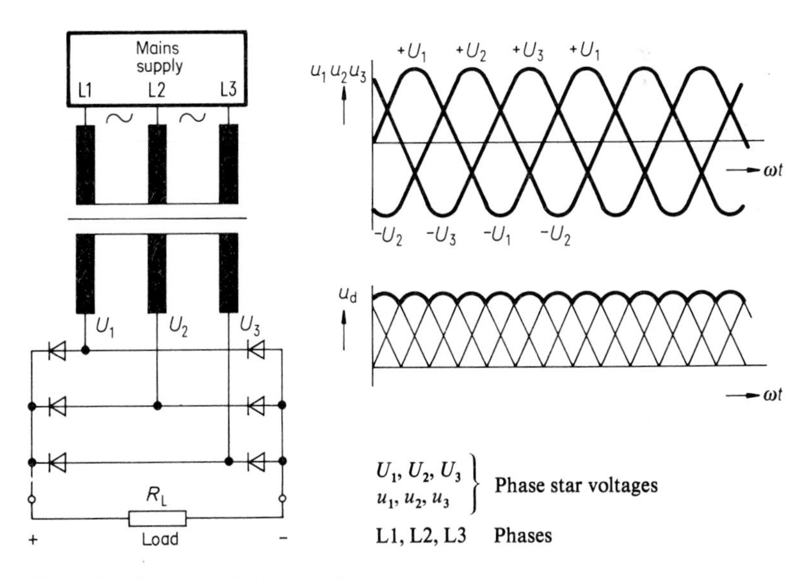

Fig. 7.15 Three-phase bridge circuit

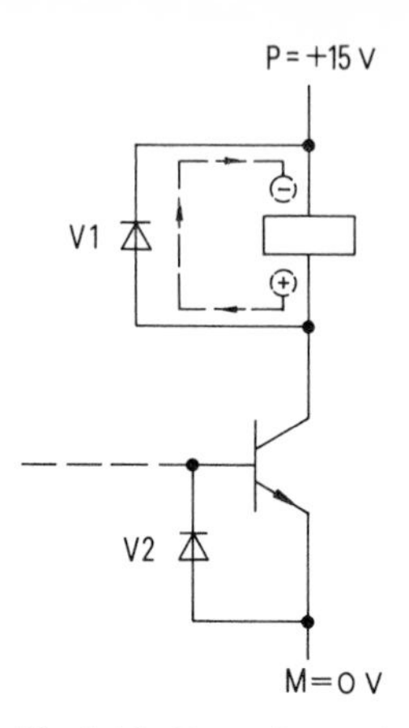

Fig. 7.16 Protection against voltage peaks

7.2.2 Protection against voltage peaks

With an arrangement as in Fig. 7.16, which is frequently encountered in practice, the magnetic flux density at the relay collapses when the transistor reverses. A voltage peak arises which is broken down by the appropriately polarized free-running diode V1 and is thus kept away from the transistor. The relay is still excited and only becomes dead after a certain time (delay time).

A further diode (V2) protects the base-emitter section of the transistor.

7.2.3 Stabilization of very small voltages

Diodes with a very steep forward characteristic are used to stabilize very small voltages (up to about 3 V). With a circuit arrangement as in Fig. 7.17 slight voltage fluctuations at the input, obtained utilizing the straight characteristic curve of the diode, do not lead to voltage fluctuations at the output. The voltage fluctuations result only in different voltages drops at the resistor R_v.

7.2.4 Reducing diode circuit

For a reducing circuit (circuit for reducing the voltage) a number of diodes are connected in series (Fig. 7.18). An undesirably high voltage, which the battery is receiving or giving off, can be reduced, e.g. by means of the overall forward voltage drop of the diodes. The reducing diodes can be activated or bridged by a reducing diode

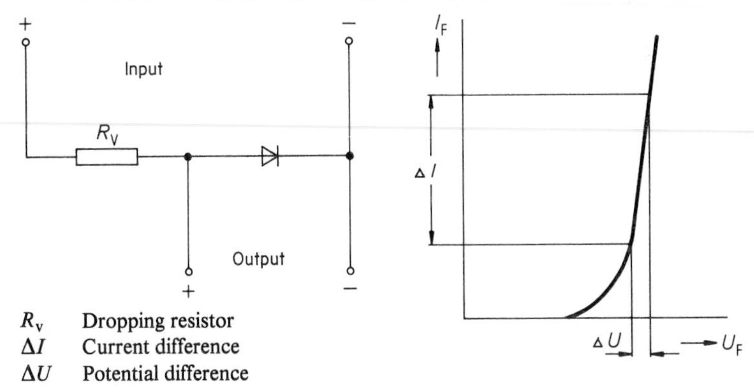

Fig. 7.17 Stabilization of voltages up to about 3 V

selector 'as a function of the voltage' using a bridging contactor. This prevents too high a voltage to the load in all operating states of the power supply system.

7.3 Silicon Zener Diode

The Zener diode (Fig. 7.19) is also called a reference or threshold diode. The *characteristic* (Fig. 7.20) shows that the Zener diode behaves like a 'normal' silicon diode in the conducting state region.

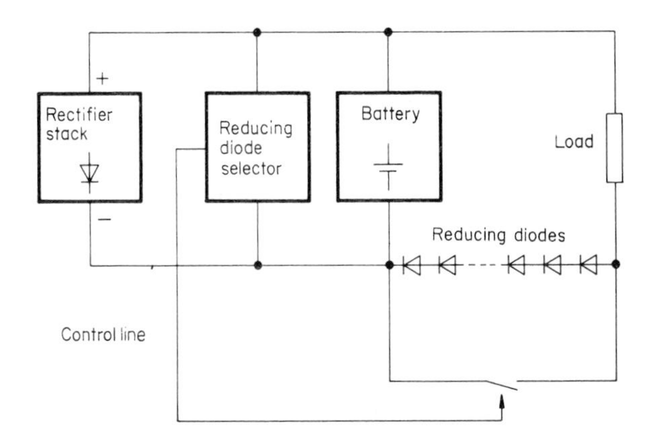

Fig. 7.18 Reducing diode circuit (counter cell circuit)

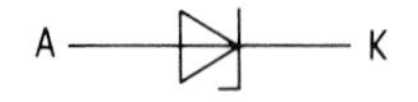

Fig. 7.19
Zener diode symbol

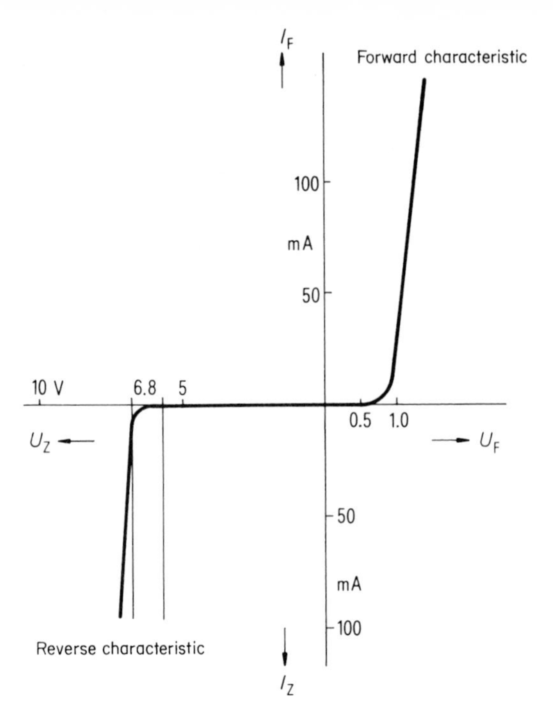

Fig. 7.20
Characteristic of a Zener diode
($U_Z = 6.8$ V)

It is its special behaviour in the reverse direction that is utilized. If a rising negative voltage is applied to the anode (as opposed to the cathode), there is only a little change in the reverse current until the breakdown voltage is reached. It then rises abruptly and very sharply, because the internal resistance of the Zener diodes becomes very small.

In the 'breakdown region' (stabilization region) the voltage at the Zener diode remains virtually constant.

The *Zener breakdown* and *avalanche breakdown* are the cause for the characteristic of the Zener diode in the reverse direction.

Zener breakdown is the name given to a phenomenon by which, after a high field strength is reached, valency electrons bonded at the p–n junction are torn out of the crystal lattice thereby becoming free electrons and holes. The number of free charge carriers increases and the internal resistance drops abruptly. The Zener breakdown is important up to about 5 V in the case of Zener diodes, whose breakdown voltage has a negative temperature coefficient. The curve of the characteristic in the

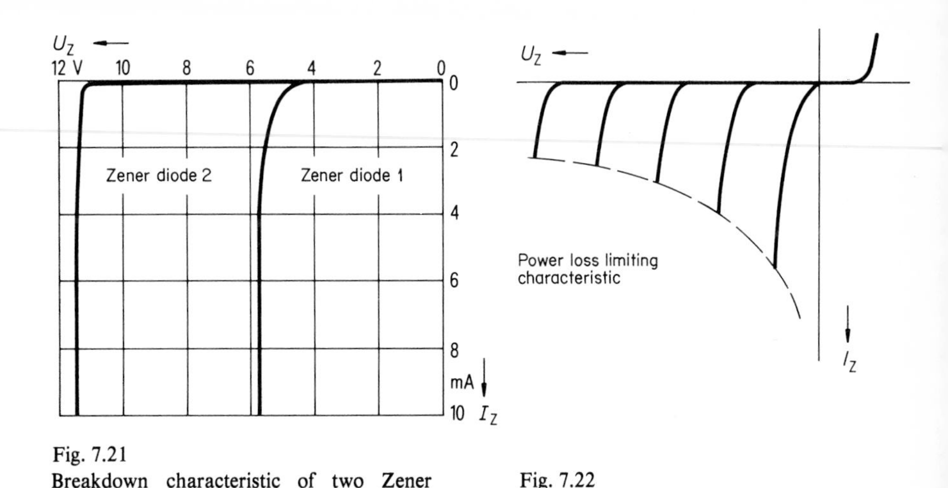

Fig. 7.21
Breakdown characteristic of two Zener diodes

Fig. 7.22
Power loss hyperbola

breakdown region is round in the case of these Zener diodes (Fig. 7.21, Zener diode 1).

Avalanche breakdown means an avalanche-type multiplication of charge carriers. The cause for this is that free charge carriers adopt such a high speed, and thereby develop energy, that valency electrons are knocked out of their bond. The number of free charge carriers thereby increases and the resistance becomes less.

The avalanche breakdown occurs at the p–n junction at an even higher field strength than with the Zener breakdown. For Zener diodes over 6 V the avalanche breakdown is the more important phenomenon. With these diodes the temperature coefficient is positive and the characteristic shows a marked curve (Fig. 7.21, Zener diode 2).

If the highest permissible power loss (Fig. 7.22) of the diode is exceeded, there is a danger of 'through alloying'.

With alternating voltage the Zener diode behaves like a filter capacitor. The dynamic resistance r_d dependent on the breakdown voltage U_Z (Fig. 7.23) is least in the case of Zener diodes with a breakdown voltage from 6 to 8 V. It is therefore often favourable to connect a number of diodes (with small voltages) in series, because the dynamic resistance of the Zener diodes connected in series is less than the dynamic resistance of *one* Zener diode.

Zener diodes are made in *versions* with glass, plastic and metal housings. They are required nowadays for rated voltages from 0.5 to 200 V and for powers up to 50 W.

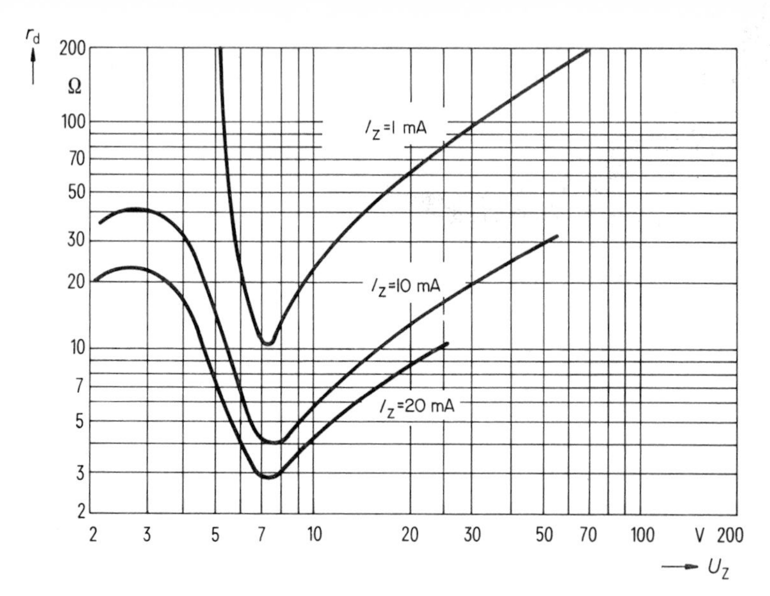

I_z Zener current

Fig. 7.23 Influence of the Zener voltage U_Z on the dynamic resistance r_d

A special form of the Zener diode is the TAZ diode[1]) (suppressor diode). It is used for overvoltage protection. Compared with the Zener diode the suppressor diode has a significantly higher pulse and surge current load capacity. The response time of about 1 ps is extremely short.

Suppressor diodes protect equipment, assemblies or sensitive electronic components against voltage peaks containing energy. They can absorb *pulse output of up to 1500 W* with max. 1 ms pulse duration.

7.3.1 Stabilization of voltages and overvoltage protection

While the arrangement described in Section 7.2.3 can be used for voltages up to about 3 V, using the Zener diode permits the *stabilization of voltages* above this range.

Figure 7.24 shows a series circuit of two Zener diodes. Here there is a voltage of 20 V at the input. The voltage drop at the Zener diodes is 13.6 V. The remaining

---

[1]) *T*ransient *A*bsorption *Z*ener diode

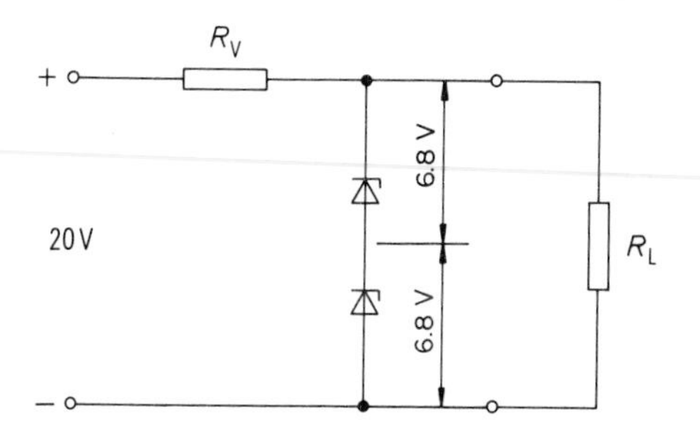

Fig. 7.24
Voltage stabilization with two 6.8 V Zener diodes

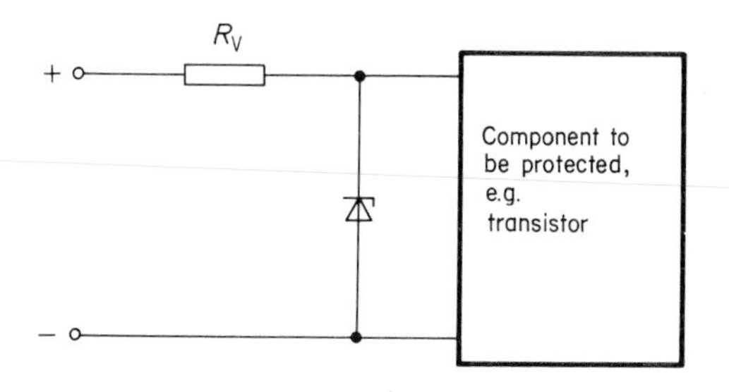

Fig. 7.25
Zener diode as overvoltage protection

voltage (6.4 V) is at the dropping resistor R_v. If the voltage at the input rises, e.g. from 20 to 22 V, the voltage at the output remains constant at 13.6 V due to the Zener diodes. The voltage drop at the dropping resistor is then 8.4 V. If the input voltage drops, e.g. from 20 to 18 V, then the voltage at the output similarly remains constant at 13.6 V (voltage drop of 4.4 V at the dropping resistor).

As Fig. 7.25 illustrates, a component (e.g. transistor) is to be *protected* against *overvoltage*. If the input voltage becomes greater than the Zener voltage, the diode passes to the conducting state. In this way any harmful overvoltage is kept away from the component.

7.4 Light-Emitting Diode

Light-emitting diode (LED), also called light diode, is the name given to a semiconductor diode which transmits light when operated in the forward direction. The colour of the light depends on the material and the doping. Gallium arsenide phosphide (GaAsP) is predominantly used for LEDs that light red. Gallium phosphide (GaP), depending on the doping of the crystalline material, leads to

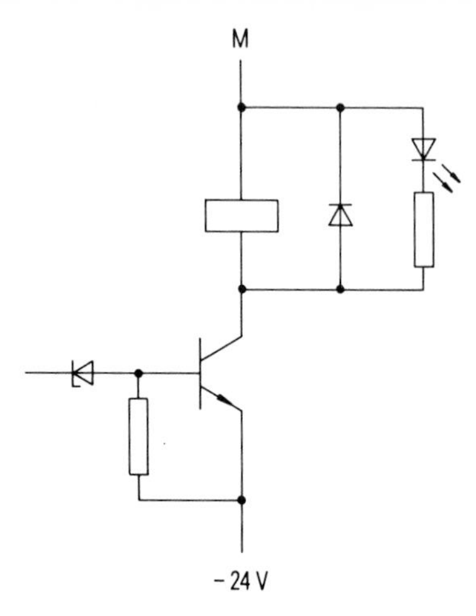

Fig. 7.26 Control circuit for LED

LEDs that light yellow or green. Gallium arsenide (GaAs) is used for infrared-emitting diodes (IRED).

The main applications of LEDs are as signal displays; IREDs are used in opto-couplers and in infrared remote control systems.

Figure 7.26 shows the example of a *control circuit* for an LED as used for voltage monitoring systems (e.g. overvoltage limiters). If the direct voltage exceeds a certain set value then, with the aid of an operational amplifier (not shown in the figure), the Zener diode and with it the transistor is made conducting. The relay is excited and switches off the rectifier which caused the overvoltage by way of the relay and contactor control system. The LED lights and indicates the 'overvoltage' state.

7.5 Silicon Transistor

There are bipolar and unipolar transistors. With the *bipolar transistor* positive defect electrons (holes) *and* free negative electrons are involved in the transport of current as charge carriers. In contrast, with the *unipolar transistor* (field effect transistor) either only positive defect electrons *or* free negative electrons are always present to transport the current as charge carriers (cf. Section 7.5.2).

The following considerations for their use as amplifiers and switches relate to the bipolar transistor.

7.5.1 Bipolar transistor

The bipolar transistor (n–p–n- or p–n–p-type) has three semiconductor regions with different doping (Fig. 7.27). The transistor has therefore in contrast to the diode two p–n junctions. The symbols and connection designations are given in Fig. 7.28.

For the transistor to pass from a *non-conductive* to a *conductive* state there must, in the case of an n–*p*–n transistor, be a positive potential at the base, in relation to the emitter. With the p–*n*–p transistor, on the other hand, it must be a negative one (Fig. 7.29).

The currents and voltages between the connections are designated as follows (Fig. 7.30):

- ▷ Emitter current I_E
- ▷ Base current I_B
- ▷ Collector current I_C
- ▷ Base-emitter voltage U_{BE}
- ▷ Collector–base voltage U_{CB}
- ▷ Collector-emitter voltage U_{CE}

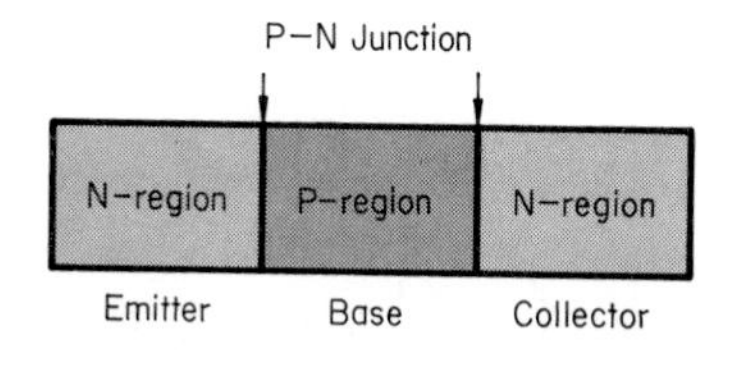

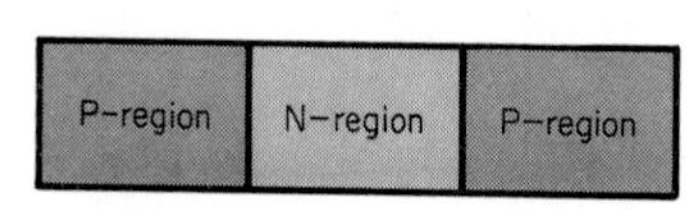

Fig. 7.27
Layer construction of the n–p–n transistor (top) and p–n–p transistor (bottom)

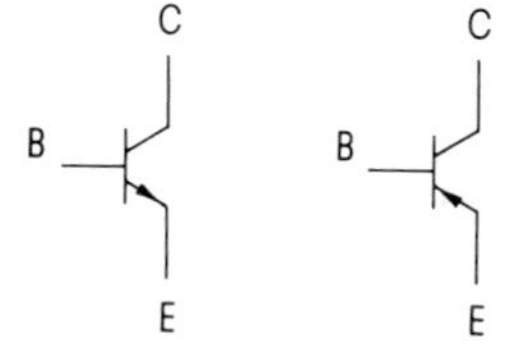

Emitter E
Base B
Collector C

Fig. 7.28
Symbols and connection designations: n–p–n (left), p–n–p (right)

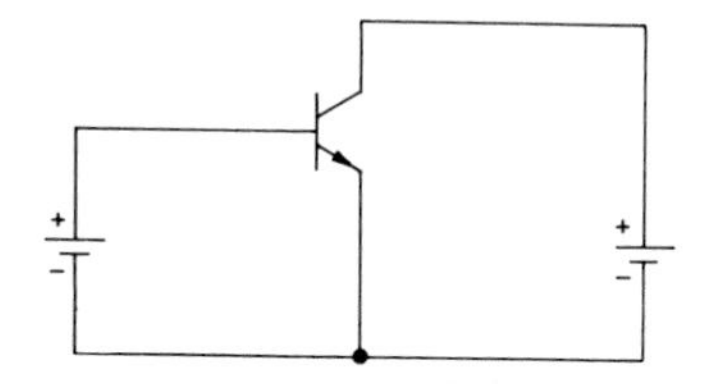

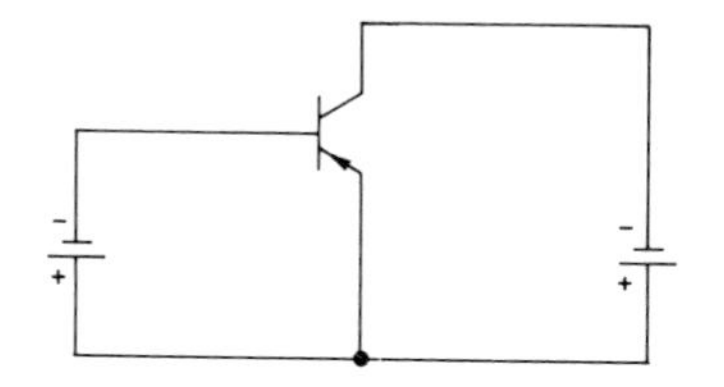

Fig. 7.29 Polarization in forward direction: n–p–n (left), p–n–p (right)

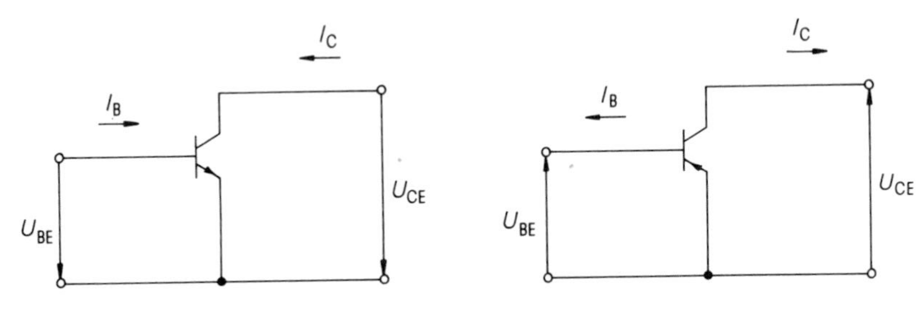

Fig. 7.30 Currents and voltages of the transistor: n–p–n (left), p–n–p (right)

The emitter current I_E breaks down into the base current I_B and the collector current I_C:

$$I_E = I_B + I_C.$$

The *input characteristic* of a transistor (Fig. 7.31) shows the connection between the base current I_B and the base-emitter voltage U_{BE}; it is similar to that of a diode operated in the forward direction.

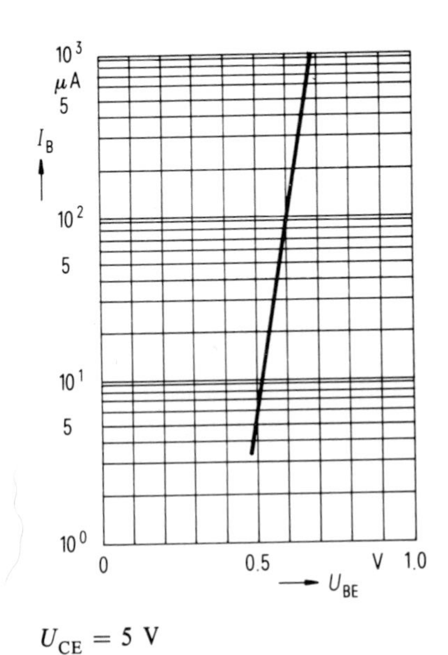

$U_{CE} = 5$ V

Fig. 7.31
Input characteristic $I_B = f(U_{BE})$ of an n–p–n silicon transistor

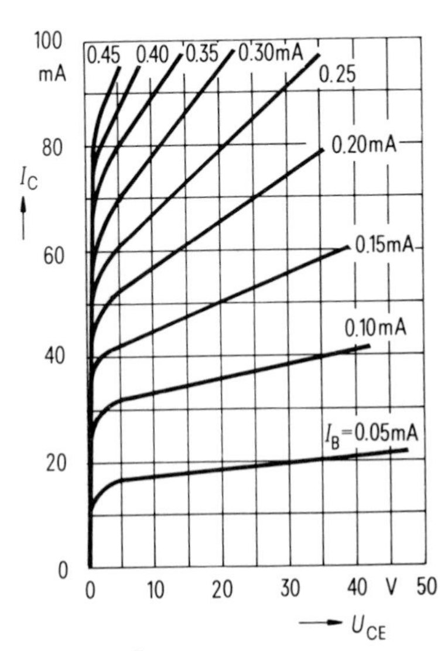

I_B Parameter

Fig. 7.32
Output characteristics $I_C = f(U_{CE})$ of an n–p–n silicon transistor

Figure 7.32 shows the *output characteristics* of a transistor. If there is no base current flowing ($I_B = 0$), the output characteristic has the same shape as the reverse characteristic of a diode. On activation of the transistor at the base, the base current I_B causes a collector current I_C which is higher by the factor of the static *current amplification* (in the emitter circuit $B = I_C/I_B$). Using the output family of characteristics it is possible to read off the relationship between the collector current I_C and the collector-emitter voltage I_{CE} as a function of the base current I_B.

Figure 7.33 illustrates *versions* of bipolar transistors.

The power loss P_{tot} heats the transistor. The heat must be dissipated in the surrounding air so that the component does not become impermissibly hot. The degree of heat dissipation is indicated by the thermal resistance R_{th}. The thermal resistance between the depletion layer (heat source) and still surrounding air (R_{thJU}) is made up of the resistance between the depletion layer and casing (R_{thJG}) and that between the casing and air. If the thermal resistance R_{thJU} in relation to the power loss P_{tot} occurring in a particular application is too high, *cooling* must be provided, e.g. by mounting the transistor on a cooling sheet or heat sink. This reduces the thermal resistance between the transistor casing and air. Dissipation is now determined by the thermal resistance between the cooling sheet and air.

Transistor as amplifier

Figure 7.34 illustrates the action of the transistor as an amplifier. This is an emitter circuit (cf. Fig. 7.35). If a very large value is set with R, then no base current I_B can flow and thus also no collector current I_C. The transistor is thus *non-conducting* and the emitter-collector space is highly resistive. The lamp does *not* light (Fig. 7.34a).

Fig. 7.33 Bipolar transistors

If a small base current is set, e.g. 0.5 mA, then the collector current would be already 50 mA. The transistor becomes *conducting* and the emitter-collector space is thereby low resistive. The lamp *lights* (Fig. 7.34b).

With a base current of, for example 1 mA, the collector current would already be 100 mA. The transistor is now *fully advanced controlled* ('even more conductive'), as in the example in Fig. 7.34(b). The resistance value for the emitter-collector space becomes even lower and the lamp lights *even brighter* (Fig. 7.34c).

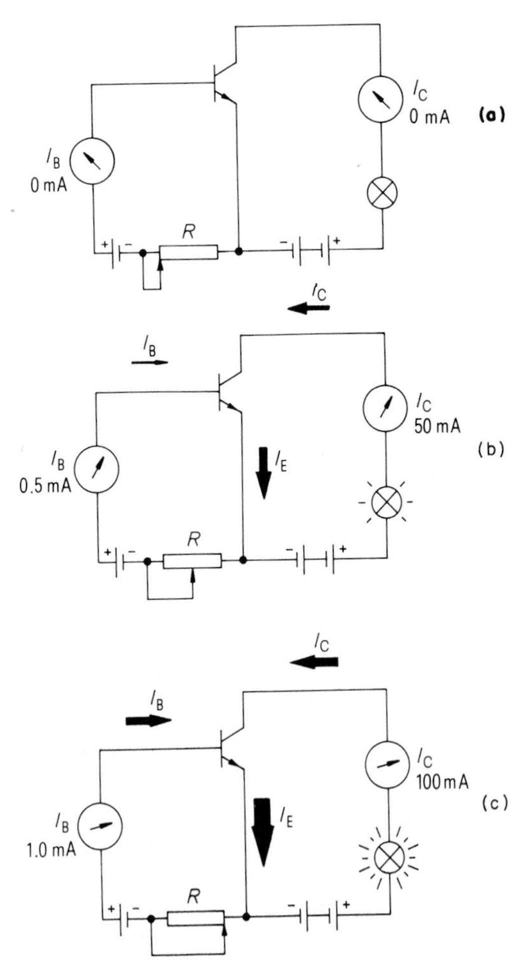

Fig. 7.34 Transistor as amplifier (emitter circuit)

In addition to the most frequently used *emitter circuit* for largest current *and* voltage amplification (which has already been described), there is also the *collector circuit* permitting a high input resistance and the *base circuit*. This latter basic circuit is used to advantage for very high frequencies (Fig. 7.35).

Transistor as switch

Figure 7.36 shows how the transistor acts as a switch. If switch S is opened

Characteristic quantities	Type of circuit		
	Emitter circuit	Base circuit	Collector circuit
	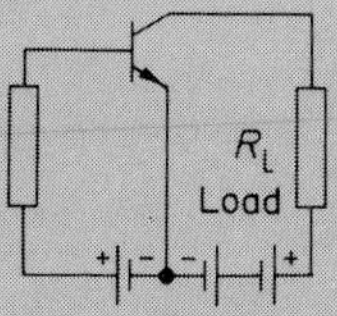 R_L Load	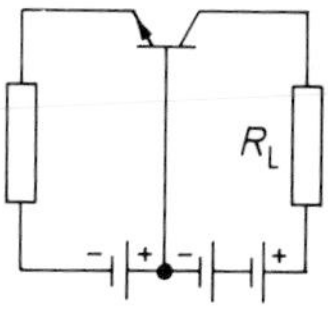 R_L	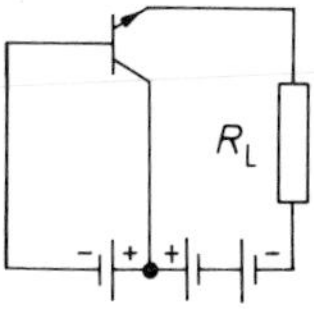 R_L
Current amplification	large 100 to 200fold	small <1	large 100 to 200fold
Voltage amplification	large 100 to 1000fold	large 100 to 1000fold	small <1
Power amplification	very large 1000 to 10,000fold	large 100 to 1000fold	medium 10 to 200fold
Input resistance	medium	very small	large
Output resistance	large	very large	very small
Limit frequency	low to medium	high	low to medium
Phase position between input and output voltage	the output signal is phase-shifted 180° against the input signal	the output signal is not phase-shifted	the output signal is not phase-shifted
Application	analog and digital amplifiers	HF amplifier	impedance converter

Fig. 7.35 Comparison of three basic transistor circuits

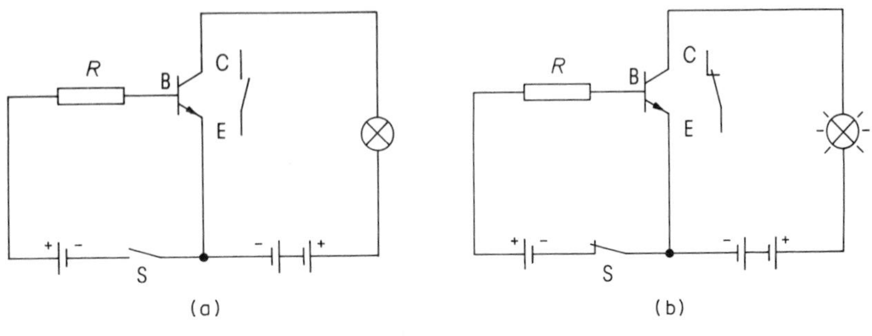

Fig. 7.36 Transistor as switch

(Fig. 7.36a), then the transistor is in the *non-conducting* state and the emitter-collector space corresponds to the *broken* contact (*switch*). The lamp does not *light*.

If the switch is closed (Fig. 7.36b) the transistor passes to the *conducting* state. If the minimum emitter-collector voltage is disregarded, then the emitter-collector space can be regarded as a *made* contact. The lamp thus now *lights*.

In practice, the transistor output pulse is deformed and delayed compared with the input pulse (Fig. 7.37).

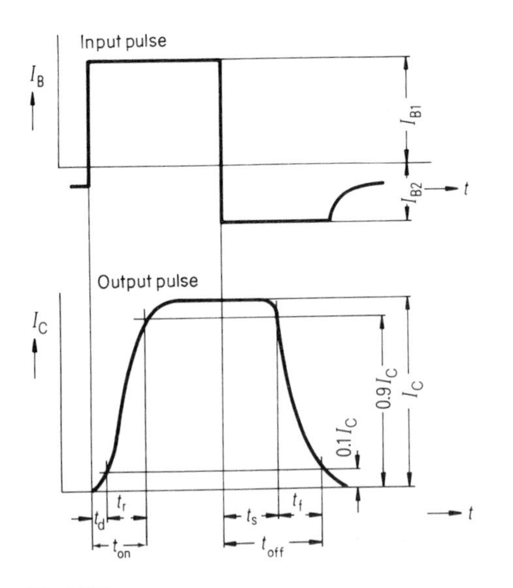

Fig. 7.37
Switching behaviour of the transistor in emitter circuit.

t_d	Delay time
t_r	Rise time
t_s	Storage time
t_f	Fall time
I_{B1}	Control current
I_{B2}	Depletion current

The *turn-on time* is the time in which the output current (collector current) rises to 90% of its maximum value after switching on the control current. It is made up of the delay time t_d and the rise time t_r:

$$t_{on} = t_d + t_r.$$

The *delay time* is defined as the time in which, after switching on the control pulse, the collector current has risen to 10% of its final value.

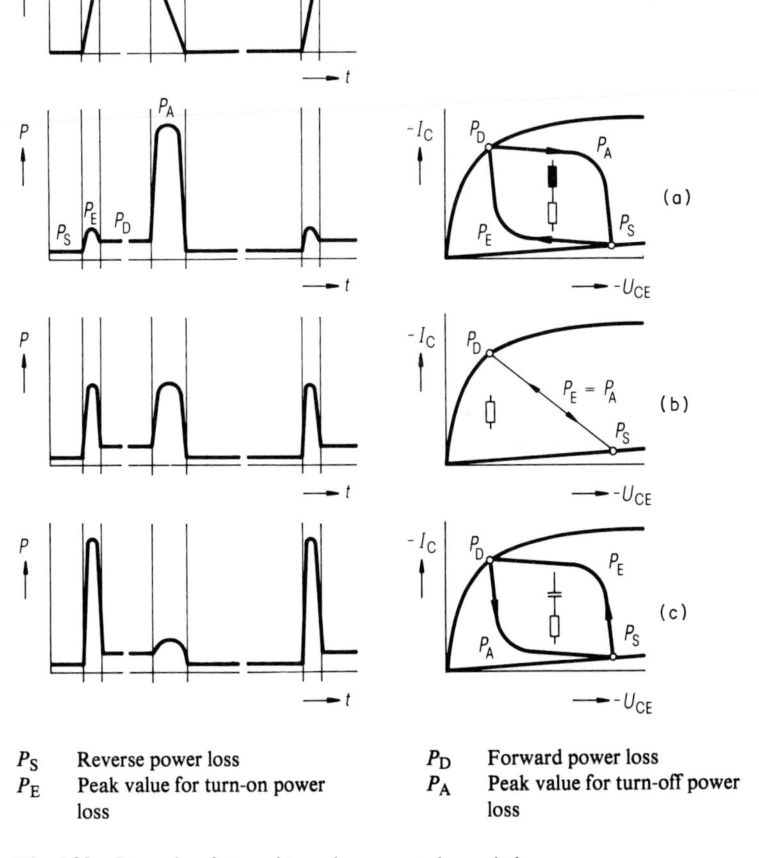

P_S	Reverse power loss	P_D	Forward power loss
P_E	Peak value for turn-on power loss	P_A	Peak value for turn-off power loss

Fig. 7.38 Power loss in transistor when operated as switch

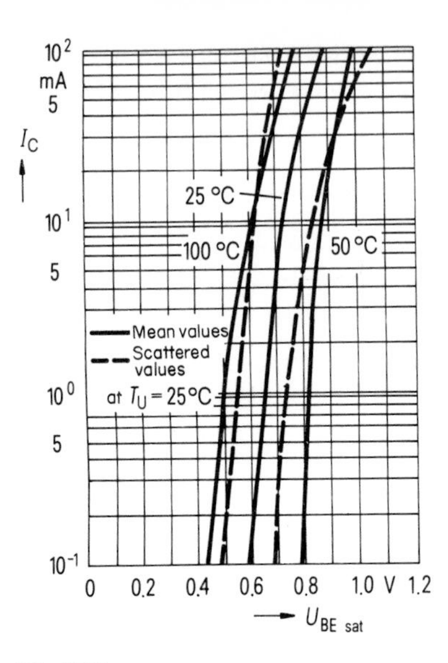

T_U Ambient temperature (parameter)

Fig. 7.39
Saturation voltage $U_{BE\,sat} = f(I_C)$ of an n–p–n silicon transistor (Static power amplification in the emitter circuit $B = 20$)

The *rise time* is that time in which the collector current rises from 10 to 90% of its final value.

Turn-off time is the term given to the time in which, after switching off the control pulse, the output current falls to 10% of its maximum value. It is made up of the storage time t_s and the fall time t_f:

$$t_{off} = t_s + t_f.$$

The *storage time* is the time in which, after switching off the control current, the output current drops to 90% of its maximum value.

Fall time is the term given to the time in which the output current falls from 90 to 10% of its maximum value.

Power loss

During a switching period the transistor is loaded by reverse, turn-on, forward and turn-off losses. Figure 7.38 shows the diagrams for the power losses occurring in the transistor for inductive (a), resistive (b) and capacitive (c) loads.

Inductive load acts (acritically) when switching on a delayed current rise. When switching off, however, the current initially endeavours to continue flowing at its full

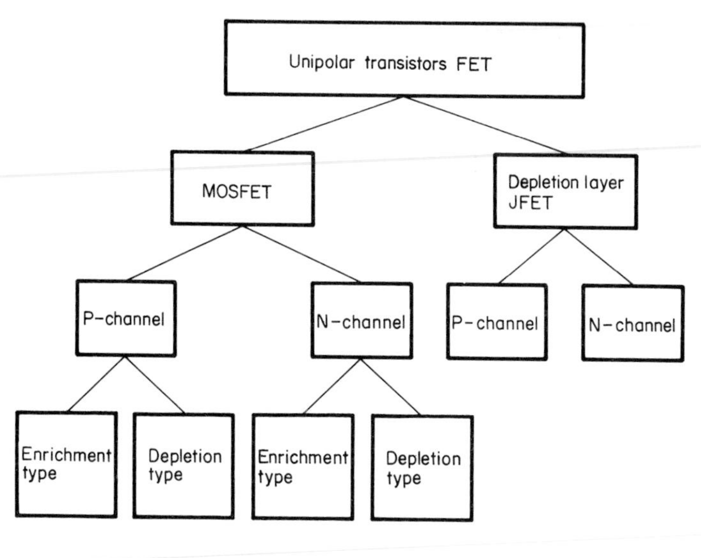

FET	Field effect transistor
MOS	Metal oxide (silicon) semiconductor
J(JG)	Insulated gate
P-channel	Positive channel (current transport through holes)
N-channel	Negative channel (current transport by electrons)

Fig. 7.40 Classification of unipolar field effect transistors (FET)

value. A reverse voltage higher than the operating voltage can thus occur due to the high reserve resistance of the transistor. For this reason a particularly high power loss occurs for a short time.

In the case of *resistive load* the operating point moves along the same load lines during the turn-on and turn-off processes. The losses are thereby equally large.

If a transistor switches on a *capacitive load*, the current is very large if the collector-emitter voltage applied is still high, resulting in a large power loss. When switching off, the current falls very rapidly, resulting in only a small power loss.

Figure 7.39 shows the connection between the collector current I_C, saturation voltage $U_{BE\,sat}$ and temperature T_U.

7.5.2 Unipolar transistor

With the unipolar transistor the number of charge carriers available in the semiconductor region, and thus the resistance of this region, is controlled by an electrical field. This arises through a voltage between a control electrode and a reference electrode.

Figure 7.40 shows how unipolar field effect transistors are classified.

Depletion layer field effect transistor

In a p-channel JFET the holes take over transport of the current while in an n-channel JFET the electrons do so. Figure 7.41 shows an example of an n-channel depletion layer field effect transistor with layer construction (a) and graphical symbol (b).

Two p-regions are diffused into the side of some n-conducting semiconductor materials. This results in two p–n junctions.

A current source (1) is connected to the ends of the n-conducting semiconductor material. The negative pole is applied at connection S (source) and the positive one at connection D (drain). Thus the electrons flow from S to D in the n-channel.

The G connection (gate) is connected with the negative pole of the power source 2. On raising the negative gate voltage the p-space charge regions due to occurrence of

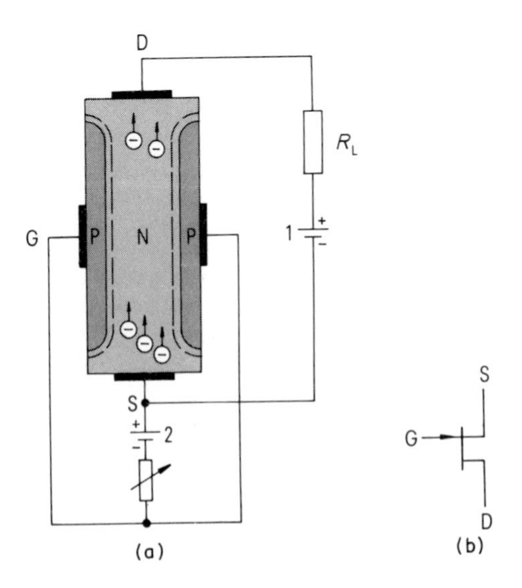

Fig. 7.41
a) Layer construction and external wiring and b) graphical symbol of the n-channel JFET

the 'depletion layer field effect' spread into the channel thereby constricting the current path. Constricting the n-channel means raising the resistance of this region, whereby the current flow is reduced.

If the negative gate voltage is reduced, then the p-space charge regions become narrower, resulting in less constriction of the n-channel. The resistance of the n-channel thereby becomes less and the current flow greater. The voltage at the gate G thus influences the current from S to D.

MOS field effect transistor

There are two types of MOS field effect transistor: p-channel MOSFET and n-channel MOSFET. Each type can be further subdivided as either of the enrichment type or depletion type (cf. Fig. 7.40).

An enrichment (or enhancement) type of MOSFET is defined as one in which the channel region is enriched with charge carriers through the gate voltage by allowing a charge carrier current to flow. With this type *no* flow of charge carrier current is possible without gate voltage.

In the depletion type the channel region is depleted of charge carriers by the gate voltage and the charge carrier current is thereby reduced. A flow of charge carrier current is possible without gate voltage.

A p-channel MOSFET enrichment type can be seen as an example in Fig. 7.42. With a negative voltage at the gate electrons are forced into the inside of the crystal and holes are drawn to the surface. A narrow p-conducting layer occurs beneath the surface (channel). Current can flow between the two p-regions, the source and drain.

Power MOS field effect transistor

SIPMOS® power transistors[1]) are n-channel field effect transistors. Figure 7.43(a) shows the layer construction and Fig. 7.43(b) the graphical symbol of this type.

There is a poor conducting n-layer on a relatively thick, though well conducting, n^+-layer. In the surface cellular n^+-conducting source regions are arranged in p-conducting troughs. A polysilicon gate electrode, insulated and embedded in quartz, covers the surface between the source cells, which are connected together by metallizing the whole surface of the source. The gate electrode has specifically profiled edges which, when making the extremely short channel region, serve as a mask for the implantation of n- and p-dopings.

---

[1]) Siemens Power MOS.

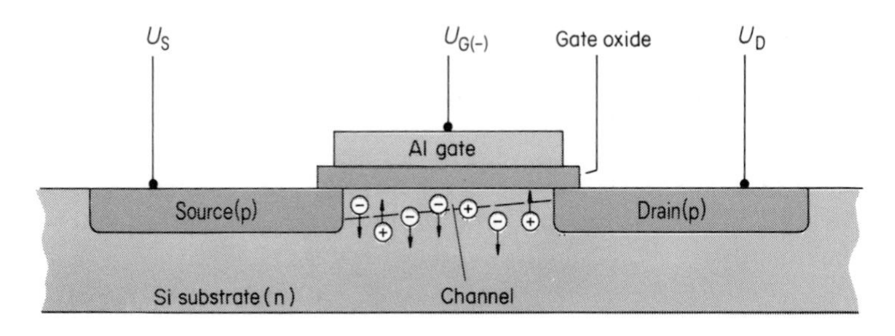

Fig. 7.42 P-channel MOS field effect transistor (enrichment type)

With a positive gate–source voltage the current under the SiO_2–Si interface is produced by electrons, drawn there by the electrical field of the gate electrode.

The advantages of the SIPMOS power transistor are primarily:

- ▷ high power amplification,
- ▷ rapid switching,
- ▷ low internal resistance,
- ▷ low control power requirement,
- ▷ problem-free connection in parallel (no second breakdown).

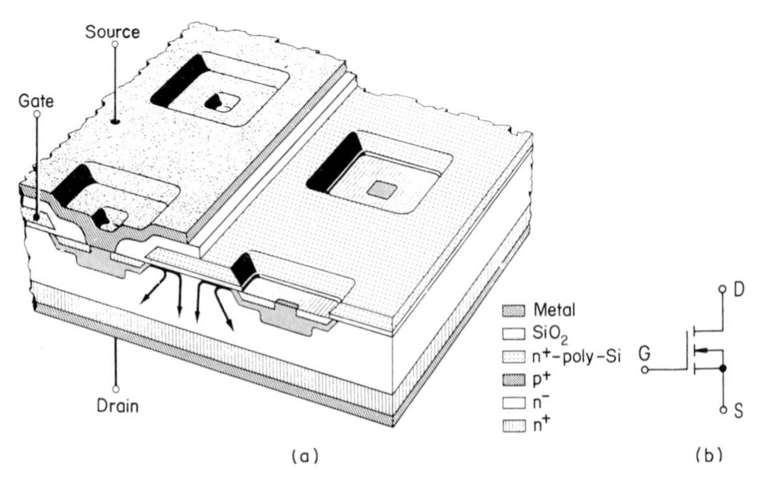

Fig. 7.43
a) Layer construction and b) graphical symbol of the SIPMOS® power transistor

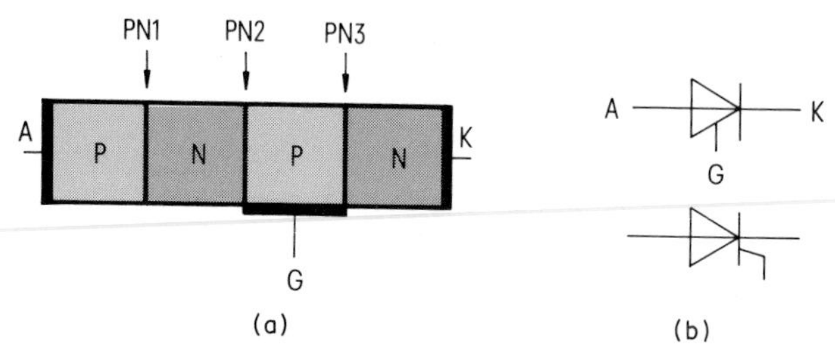

G Control gate

Fig. 7.44 Thyristor: a) layer construction and b) symbols

SIPMOS transistors are used by preference as rapid-action switches in power electronics, e.g. for switching mode power supplies and converters.

7.6 Silicon Thyristor

The thyristor, one of the most important components in telecommunications power supply equipment, consists of four layers with p–n–p–n doping and three p–n junctions.

In contrast to the diode the thyristor is controllable, a gate being used for this purpose (Fig. 7.44).

7.6.1 Working and characteristics

Figures 7.45, 7.46 and 7.47 show the working of the thyristor.

Reverse off state

In Fig. 7.45(a) the anode is connected to the negative pole of battery 2 and the cathode to the positive pole. The p–n junctions PN1 and PN3 are highly resistive, while the p–n junction PN2 is of low resistivity. The thyristor is thus non-conducting. Regardless of whether switch S1 is open or closed, the thyristor cannot be brought from the reverse off state to a conducting state until the poles of battery 2 are reversed. Figure 7.45(b) replaces the layer construction with the graphical symbol. The lamp *cannot* light because the anode–cathode space acts like an open switch.

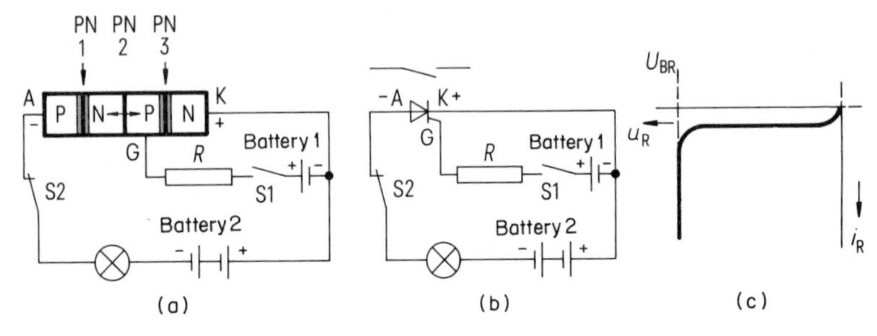

u_r Reverse or negative off voltage
i_R Reverse or negative off current
U_{BR} Reverse or negative breakdown voltage

Fig. 7.45 Reverse off state of the thyristor

Figure 7.45(c) shows the reverse off characteristic (negative off characteristic). Below the 'breakdown voltage' the reverse off current is largely independent of the voltage applied (reverse current level). On exceeding the breakdown voltage the reverse current rises steeply and the component can be destroyed (steep reverse current rise).

Forward off state

Compared with Fig. 7.45 the poles of battery 2 have been reversed (Fig. 7.46a). The positive pole of battery 2 is thus now connected to the anode and the negative pole

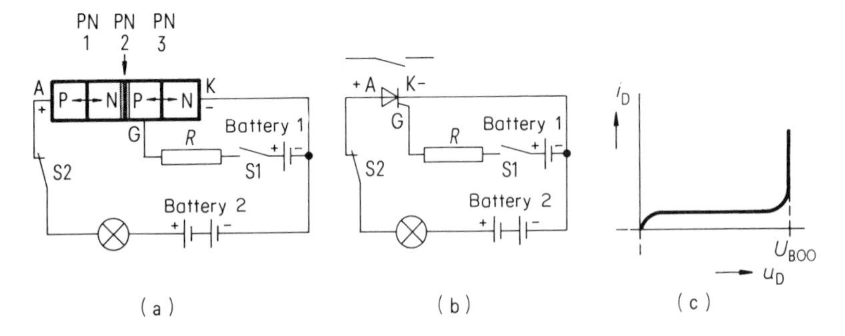

U_{BOO} Zero breakover voltage
i_D Forward or positive off current
u_D Forward or positive off voltage

Fig. 7.46 Forward off state of the thyristor

to the cathode. Switch S1 is still open. Now the p–n junctions PN1 and PN3 have low resistivity, while the p–n junction PN2 is highly resistive. The anode–cathode space again corresponds to an open switch (Fig. 7.46b). The lamp again does *not* light (cf. Fig. 7.45a and b).

Figure 7.46(c) shows the forward off characteristic (positive off characteristic). The forward off current in wide ranges is independent of the voltage applied.

If the 'zero breakover voltage' is exceeded the thyristor will fire (undesired effect). The zero breakover voltage usually lies around the same level as the breakdown voltage (cf. Fig. 7.46c with Fig. 7.45c).

Forward on state

Switch S1 is closed (Fig. 7.47a), producing a *control circuit* from the positive terminal of battery 1 via switch S1 and resistor R to gate G, on via the p-region and the p–n junction (PN3), then via the n-region to the cathode and from there back to the negative terminal of battery 1. The p–n junction PN2 thereby also becomes of low resistivity so that now *all* p–n junctions are of low resistivity. The thyristor is fired and conducts (Fig. 7.47a, b, c).

As the anode–cathode space corresponds to a closed switch, the result is the *main circuit* from the positive terminal of battery 2, lamp, switch S2, thyristor/anode via all p- and n-regions and all p–n junctions to the cathode, resistor R to the negative terminal of battery 2. The lamp *lights*. Only a small voltage drop now occurs still at the thyristor. This is called, as in the case of the diode, *forward voltage*.

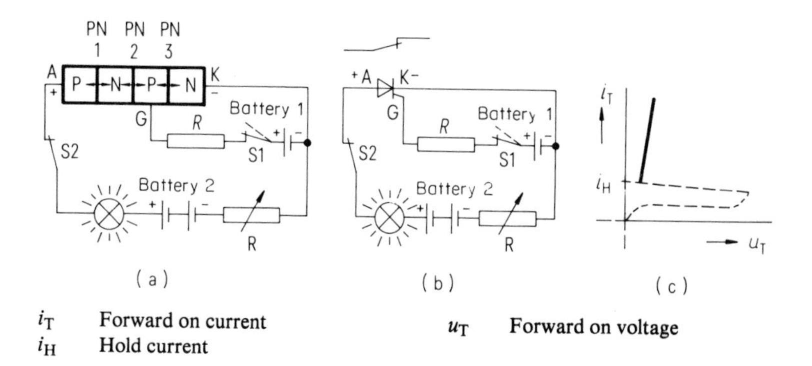

i_T Forward on current
i_H Hold current
u_T Forward on voltage

Fig. 7.47 Forward on state of the thyristor

The thyristor *remains* conducting (lamp still lit), even if the switch S1 is opened again (shown by the broken line), as long as there is no fall in the main circuit below the thyristor's *hold current* I_H. To switch the thyristor from the conducting to non-conducting state the current can be brought below the value of the hold current either by interrupting the main circuit (opening switch S2) or reducing the current in the main circuit to below the value for the hold current with the adjustable resistor R. The lamp no longer lights (cf. Fig. 7.46).

In practice the thyristor is switched off by reducing the current below the value of the hold current. If the thyristor is used in an alternating circuit, there is always automatically a fall below the hold current on each current cycle as it passes through zero.

Figure 7.48 illustrates the whole *family of characteristics* of the thyristor.

7.6.2 Designs and cooling

Thyristors for current values <20 A are made with a plastic or metal housing. For larger current values there are power thyristors with metal ceramic housings of:

screw design,
flat-bottom design,
disk design.

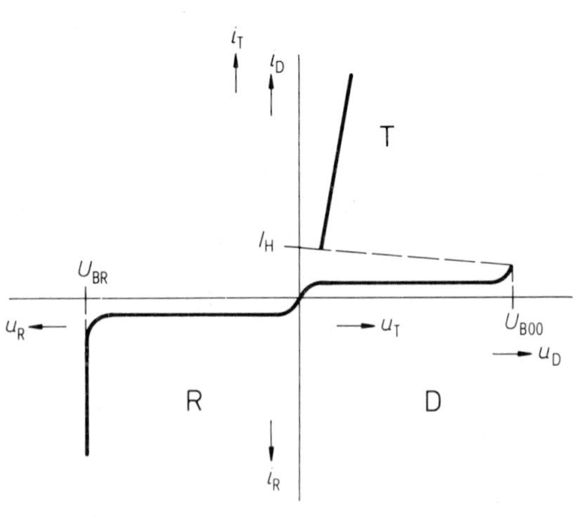

R	Reverse or negative off characteristic
u_R	Reverse or negative off voltage
U_{BR}	Reverse or negative breakdown voltage
i_R	Reverse or negative off current
D	Forward or positive off characteristic
u_D	Forward or positive off voltage
U_{B00}	Zero breakover voltage
i_D	Forward or positive off current
T	Forward on characteristic
u_T	Forward on voltage
i_T	Forward on current
I_H	Hold current

Fig. 7.48
Current and voltage characteristics of the thyristor

For some time thyristors of modular design have also been available. Figures 7.49 to 7.52 review the range of types.

Figure 7.53 shows a pressure-contacted power thyristor as an example of the *construction*. The screw pin is used for screwing it into a heat sink. Usually the screw pin forms the anode A and the flexible thick connected lead the cathode K.

As regards *cooling* a distinction is made between:

- ▷ air self-cooling,
- ▷ forced air cooling,
- ▷ liquid cooling.

Air self-cooling is normally used in a telecommunications power supply system.

A number of thyristors can be combined into a *thyristor set* (Fig. 7.54). The figure shows a thyristor set in a fully controlled three-phase bridge circuit. The thyristors (screw design) are screwed into the heat sink for air self-cooling.

Power thyristors are built in accordance with the present state of the art for continuous limiting currents of up to 3000 A, for periodic peak reverse voltages of up to 4200 V and for continuously permissible depletion layer temperatures of up to 140 °C.

7.6.3 Phase-angle control

In phase-angle control, which is very important in practice, the onset in time of the trigger pulses (control pulses, firing pulses) is shifted in relation to the sine half-wave. Phase-angle control can be explained with the aid of the single-pulse one-way circuit (Fig. 7.55). Multi-pulse circuits work in a similar way. These are described further below.

The thyristor is supplied from an alternating current source. If at a certain point in time the anode is negative compared with the cathode, then the thyristor is non-conducting. As is well known, the firing process can only be initiated if the anode is positive compared with the cathode. The *trigger set* supplies pulses for this. The positive half-waves are more or less gated, depending on the trigger pulses shifted by the control angle α (firing delay angle). In this way it is possible to regulate the voltage at the load or keep it constant.

The control angle α is calculated from the 'natural' firing time of the thyristor. This angle is measured from the point at which the current waveform passes through zero to the point at which, in a rectifier circuit with diodes, the next diode takes over carrying the current. A trigger pulse at this point produces the highest possible direct voltage at the output of the rectifier – hence the term 'full rectifier modulation' ($\alpha = 0°$).

Fig. 7.49 Small thyristors

Fig. 7.50 Power thyristors of screw and flat-bottom design

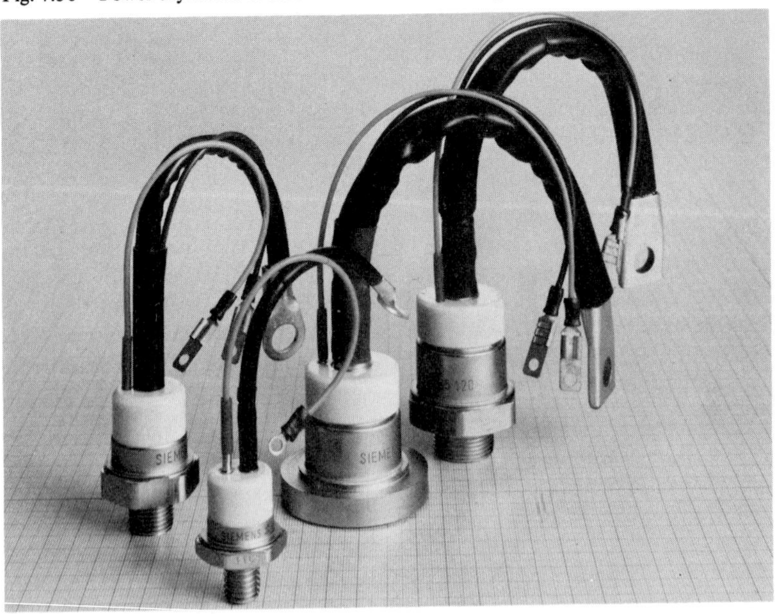

Fig. 7.51 Power thyristors of disk design

Fig. 7.52 Thyristors of modular design

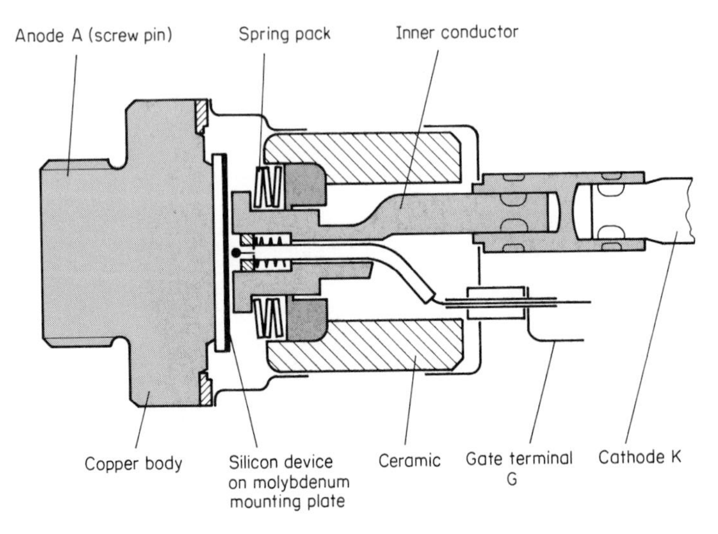

Fig. 7.53
Diagrammatic representation of a pressure-contacted power thyristor (screw design)

Fig. 7.54 Thyristor set

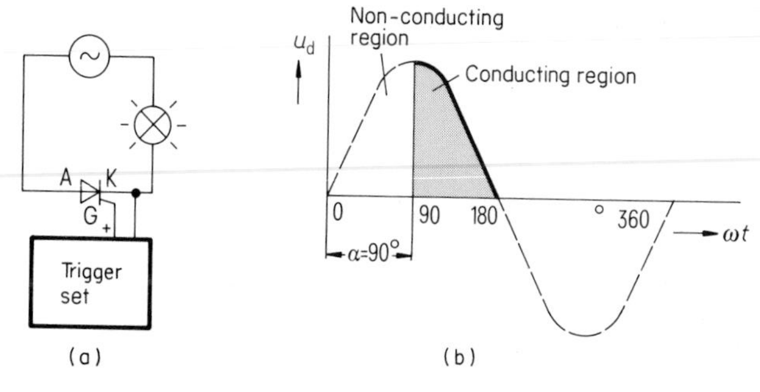

Fig. 7.55 Thyristor in single-pulse one-way circuit

Figure 7.55(b) shows a control angle of 90°. Within the range from 0° to 90° the thyristor is non-conducting and from 90° to 180° conducting.

Figure 7.56(a) shows a relatively early onset of the trigger pulse (α small). The earlier the onset, the greater the voltage–time surface becomes. A large

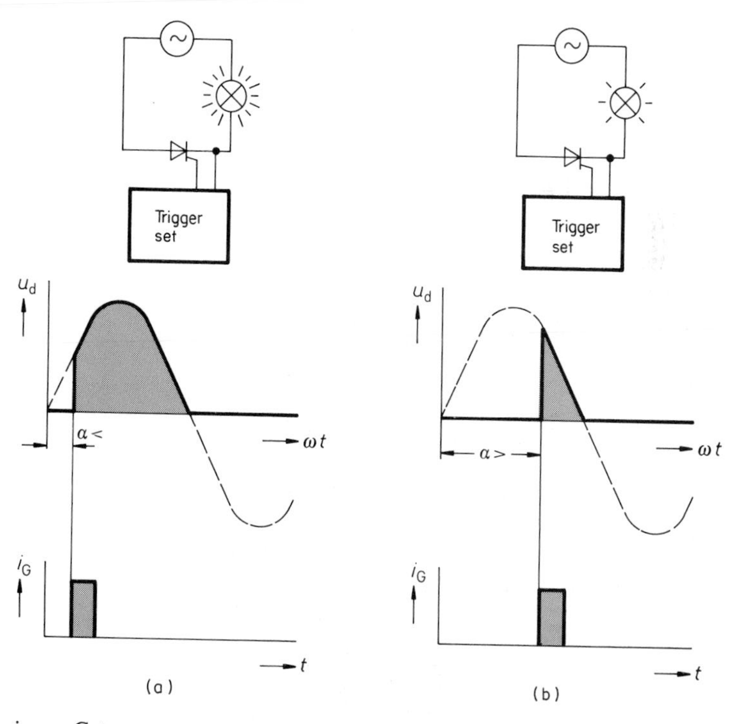

i_G Gate current

Fig. 7.56 Time shift of trigger pulses

voltage–time surface means a high voltage, so the lamp lights brightly. A larger control angle means a smaller voltage–time surface (Fig. 7.56b). If no trigger pulse reaches the gate, the thyristor remains non-conducting. The lamp accordingly does not light.

Figure 7.57 shows, with reference to Figs 7.55 and 7.56, the phase-angle control with three control angles, namely $\alpha = 45°$, $\alpha = 90°$ and $\alpha = 135°$. This relates to two periods (Fig. 7.57a) of the input a.c. voltage. Figure 7.57(b) to (d) clarifies the connections between certain control angles and the available direct voltages u_d.

The *trigger pulses*, which pass from the trigger set via the pulse transformer[1]) (Fig. 7.58e) to the thyristors of the thyristor set, must have a certain shape. This is as spikes (Fig. 7.58a), rectangular pulses (Fig. 7.58b) or 'combined pulses' (Fig. 7.58c).

With thyristor-controlled rectifiers each trigger pulse is normally cycled (7-kHz cycle, Fig. 7.58d). An advantage here is that the pulse transformer can be designed for a particularly low power.

Figure 7.58(f) shows an ideal trigger pulse. If the gate current lies within the hatched area, the thyristor can *certainly* be fired.

In practice a distinction is made with the trigger pulses used in thyristor-controlled rectifiers between 70° overlapping trigger pulses, 12° double short trigger pulses and $180° - \alpha$ overlapping long trigger pulses.

70° overlapping trigger pulses

Each trigger pulse at a frequency of 50 Hz lasts $70° \mathrel{\hat{=}} 3.88$ ms. As the trigger pulses are shifted 60°, there is an overlap of 10° (Fig. 7.59). Provided that when switching on two thyristors are simultaneously conducting, the converter is able to start handling the current. 70° trigger pulses were frequently used in equipment with a fully controlled three-phase bridge circuit.

12° double short trigger pulses

When delivering double short pulses of 12° ($\mathrel{\hat{=}}$ at 50 Hz $\approx$ 700 µs) a thyristor always receives after the 'main pulse' a second pulse shifted by 60° ('auxiliary pulse', Fig. 7.60). The main pulse for one thyristor is at the same time the auxiliary pulse for the thyristor that received its main pulse 60° beforehand. This is called *pulse overcoupling*. Because when switching on the rectifier two thyristors receive a trigger pulse simultaneously, the converter starts carrying current.

[1]) The pulse transformer is used for electrical separation and adaptation of the trigger set to the thyristors.

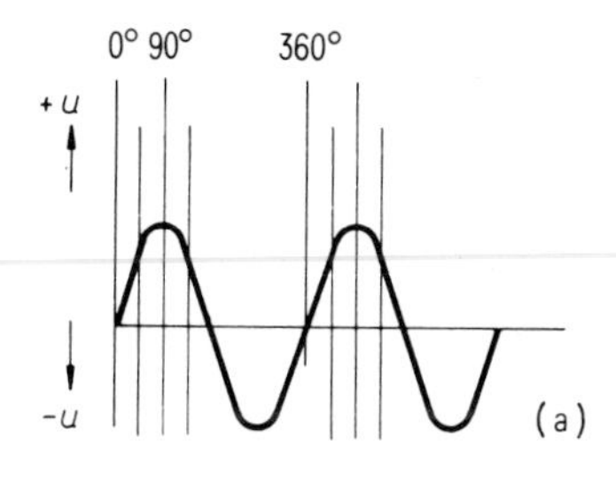

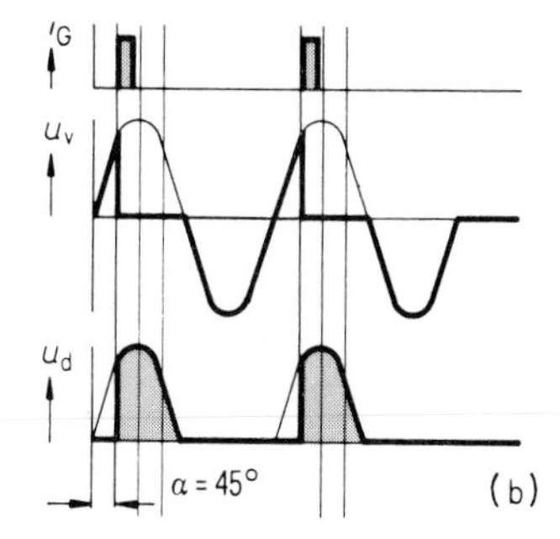

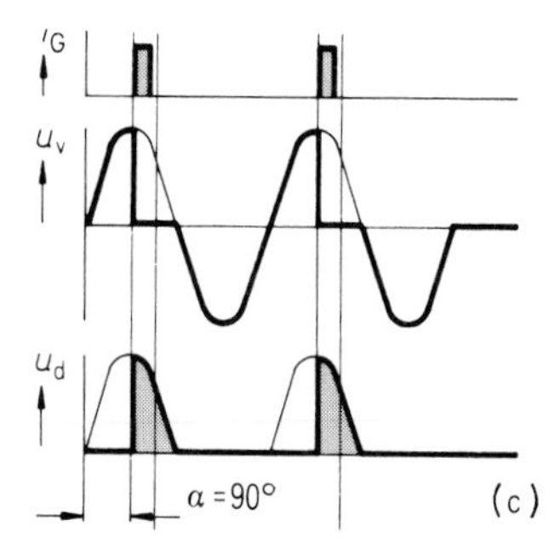

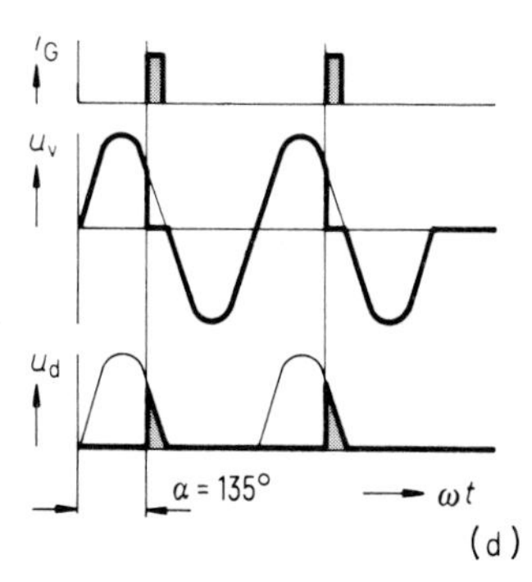

Fig. 7.57
Phase-angle control with the control angles $\alpha = 45°$, $\alpha = 90°$ and $\alpha = 135°$

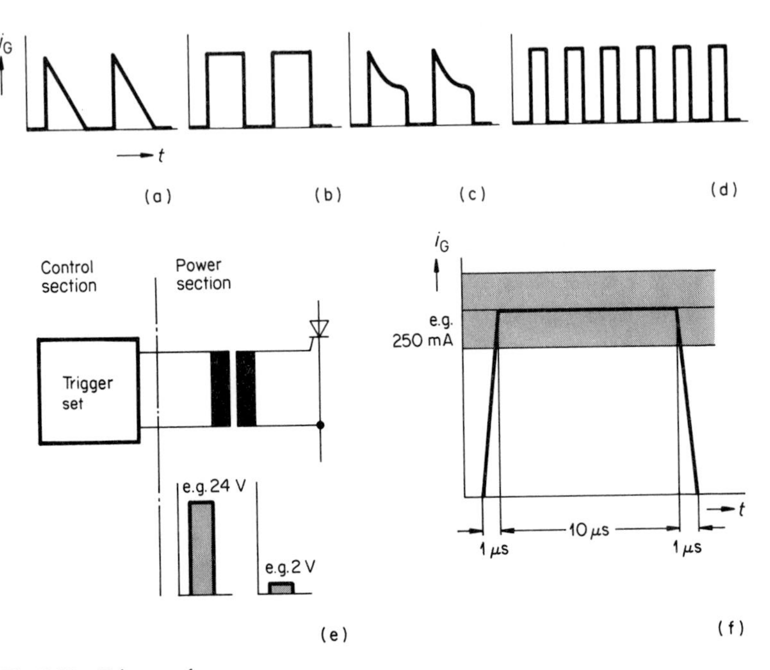

Fig. 7.58 Trigger pulses

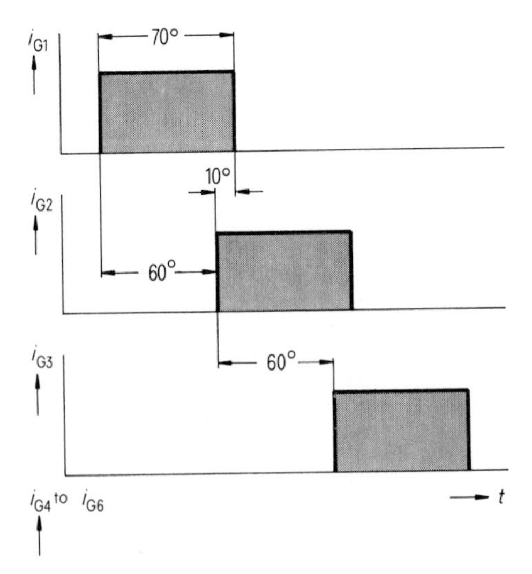

Fig. 7.59 70° overlapping trigger pulses

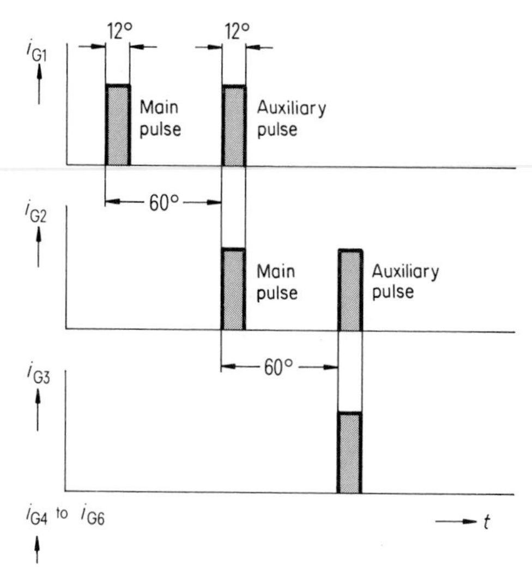

Fig. 7.60 12° double short trigger pulses

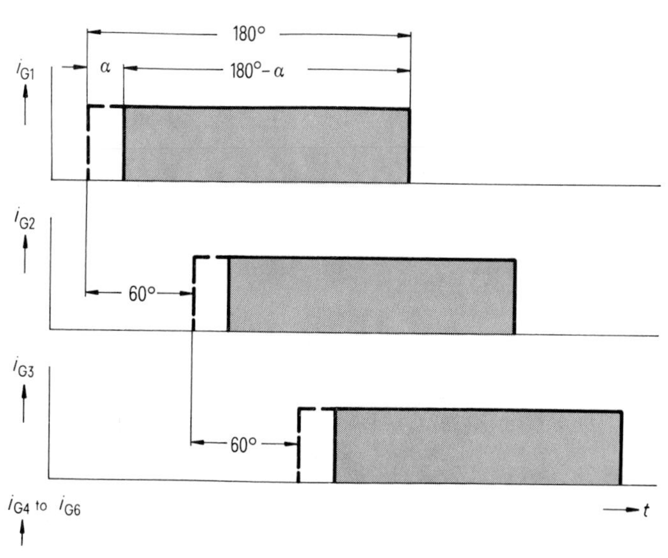

Fig. 7.61 180° – α overlapping long trigger pulses

12° double trigger pulses can also be used in equipment with a fully controlled three-phase bridge circuit.

180° – α overlapping long trigger pulses

The trigger pulses have a maximum duration of 180° (at 50 Hz, 10 ms); see Fig. 7.61. Pulse overcoupling is used here in addition to pulse overlapping (cf. Fig. 7.60).

180° – α pulses are used in the fully controlled three-phase bridge circuit, the fully controlled three-phase d.c. controller circuit and in single-phase equipment with a semi-controlled bridge circuit. In the latter equipment the trigger pulses have a phase shift of 180° (cf. Fig. 7.62).

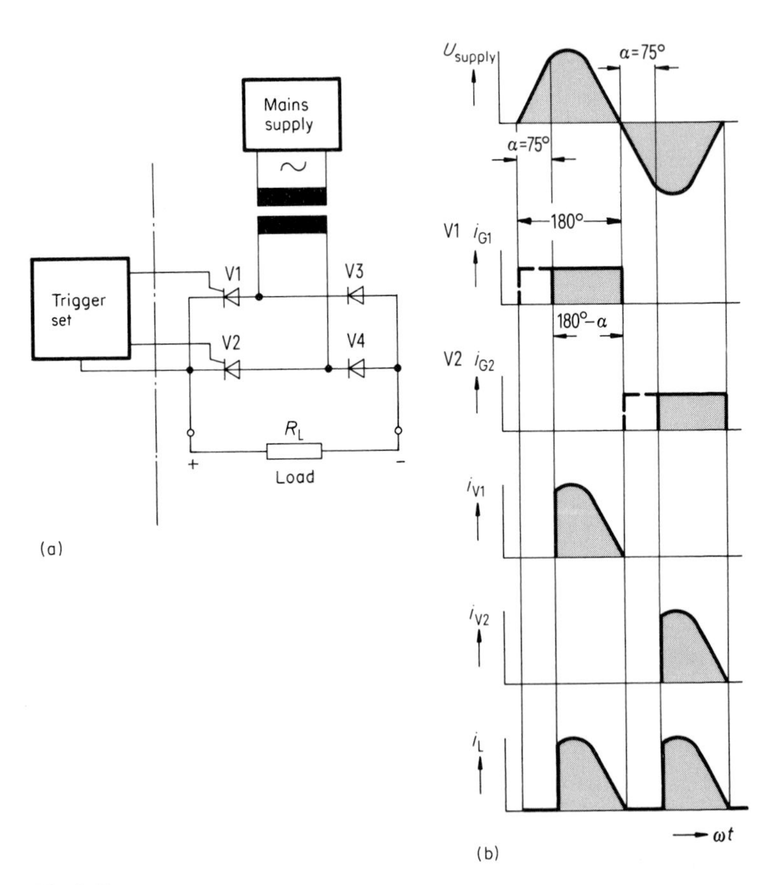

Fig. 7.62
Semi-controlled single-phase bridge circuit

Breakdown of the trigger pulses into a 7-kHz cycle is not shown in Figs 7.60 and 7.61 (cf. Fig. 7.58d).

7.6.4 Basic circuits with thyristors

Because of the importance of thyristors in the supply of power for telecommunications there now follows a detailed explanation of basic circuits designed with thyristors.

Semi-controlled single-phase bridge circuit

Two thyristors and two diodes are required in the semi-controlled single-phase bridge circuit (two-pulse circuit, Fig. 7.62). It is normally used for rectifiers up to about 25 A.

The trigger set delivers two trigger pulses per period, one for each half-wave. They are mutually phase-shifted 180° and have a duration of $180° - \alpha$. *Both* half-waves are rectified and *phase-angled*, depending on the trigger pulses (example: $\alpha = 75°$, trigger pulses $180° - 75° = 105°$). The circuit acts as a rectifier and at the same time as a final control element.

When a positive trigger pulse arrives at the gate, the following circuit is produced after firing of the thyristor V1:

> transformer (secondary side)/thyristor V1 (anode)/cathode V1/load/diode V4 (anode)/cathode V4/transformer.

Thyristor V2 can be fired after the zero passage. Thyristor V1 is now in the off state as the anode has a negative voltage compared with the cathode. When thyristor V2 receives a trigger pulse and becomes conducting, the following circuit is produced:

> transformer (secondary side)/thyristor V2 (anode)/cathode V2/load/diode V3 (anode)/cathode V3/transformer.

Figure 7.63 illustrates the curve $U_{di\alpha}/U_{di}$ as a function of the control angle α. When $\alpha = 0°$ the d.c. output voltage has the greatest value, while at $\alpha = 180°$ it is 0 V.

Fully controlled three-phase bridge circuit

The fully controlled three-phase bridge circuit requires six thyristors (Fig. 7.64). It is used for medium-sized rectifiers (e.g. 25, 40, 50, 100 and 200 A) with three-phase alternating voltage.

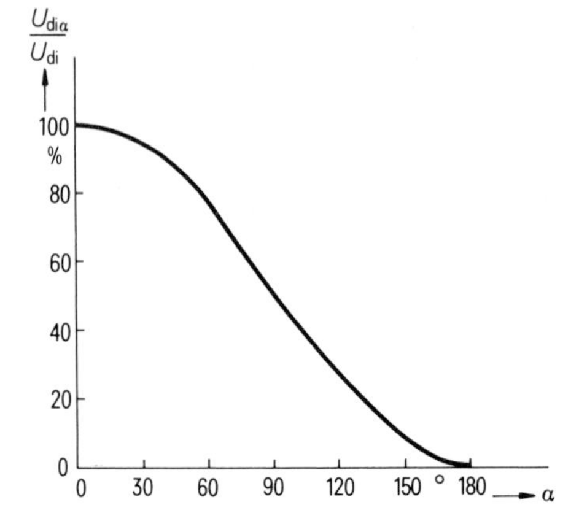

$U_{di\alpha}$ Ideal no-load d.c. voltage
U_{di} Ideal no-load d.c. voltage at $\alpha = 0°$

Fig. 7.63
Control characteristic of the semi-controlled single-phase bridge circuit

The higher frequency of the superimposed alternating voltage (300 Hz) compared with the single-phase bridge circuit permits the use of smaller filters. Also the loading of the mains by harmonics is normally reduced.

The three-phase alternating voltage reaches the thyristors V1 to V6 after step-down transformation (Fig. 7.64a).

The trigger set must deliver six trigger pulses at intervals of 60° for each supply cycle. The trigger pulses have a duration of $180° - \alpha$. The sequence in which

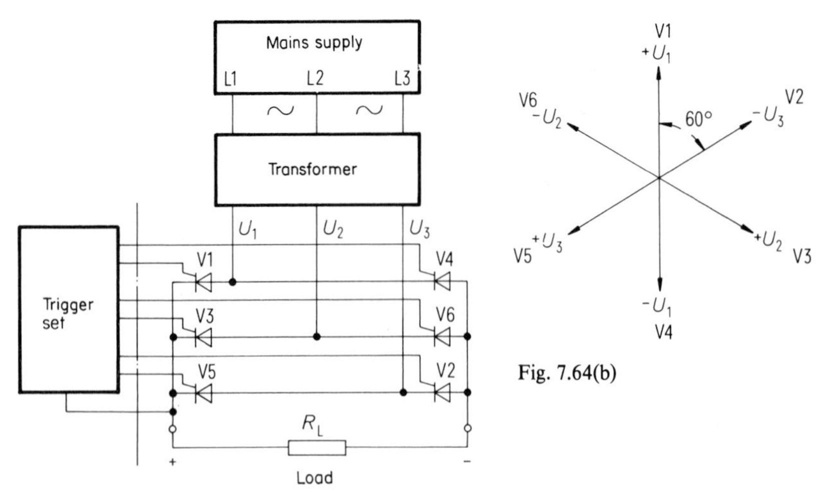

Fig. 7.64(b)

Fig. 7.64(a) Fully controlled three-phase bridge circuit

thyristors V1 to V6 must receive their pulses from the trigger set is derived from the firing sequence and arrangement of the thyristors. The circuit combines rectification and the final control element function (cf. semi-controlled single-phase bridge circuit).

Figure 7.64(b) shows the assignment of the individual voltages to thyristors V1 to V6.

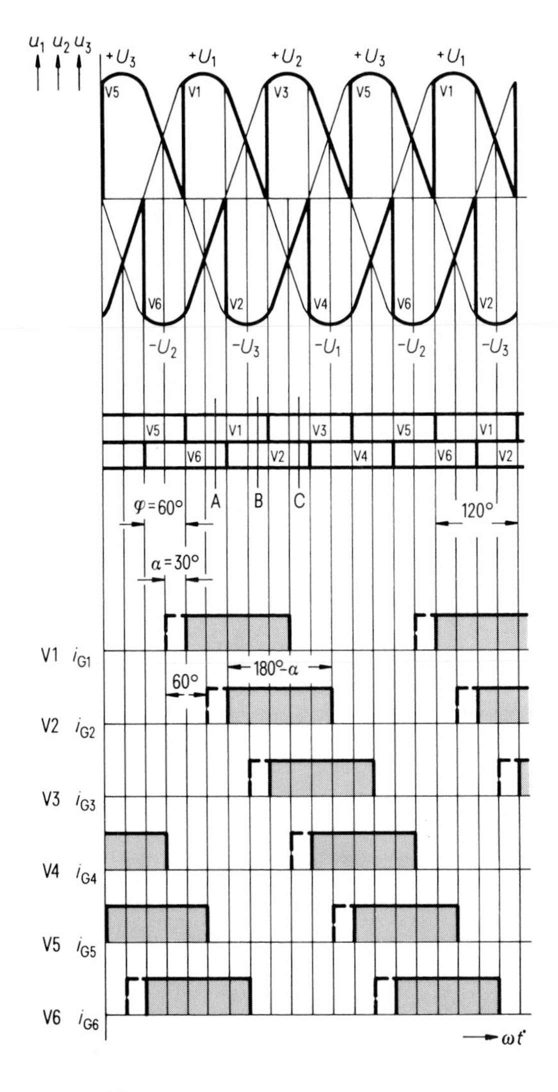

φ Phase angle

Fig. 7.64(c)

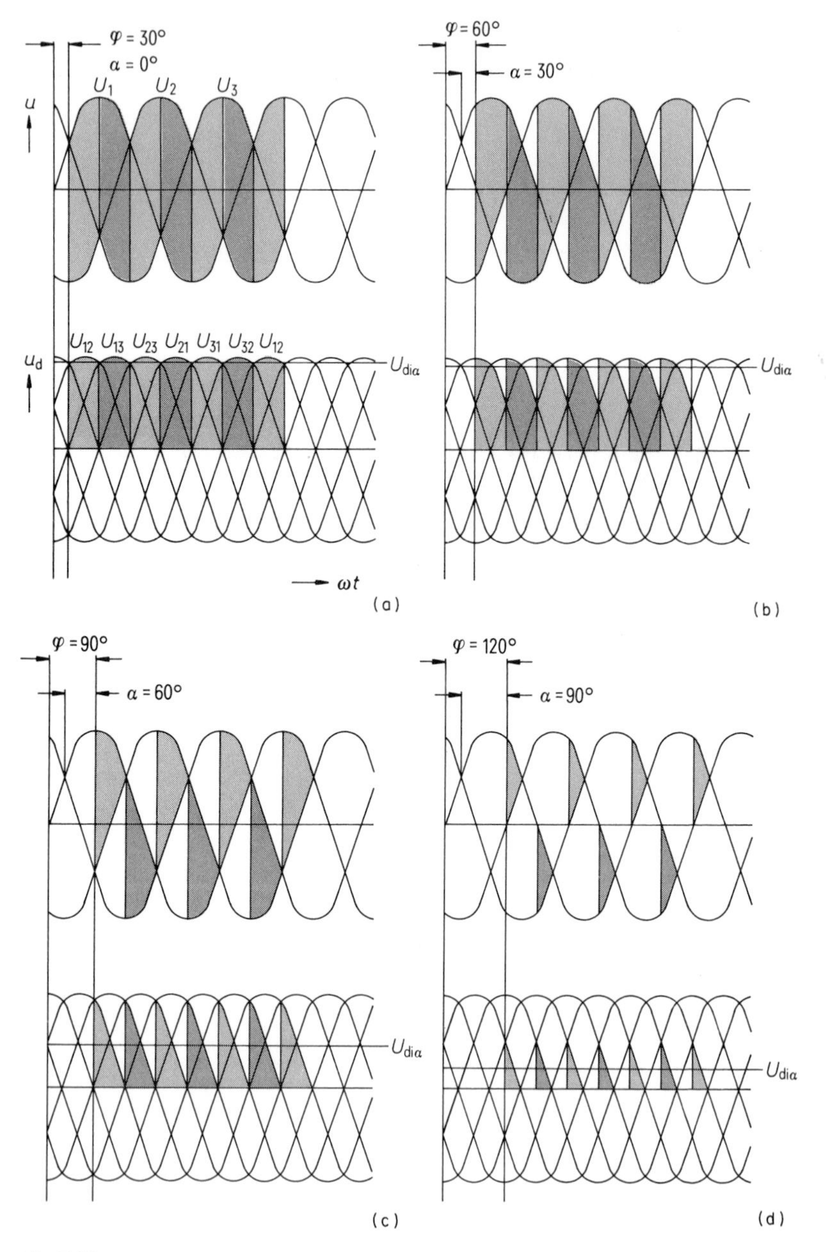

Fig. 7.65
Fully controlled three-phase bridge circuit: representation of phase-angle control and rectification at different control angles α

Figure 7.64(c) shows as an example the phase shift with a control angle of 30°.

Simultaneous firing of thyristors V6 and V1 (time A) produces the following circuit:

U_1/thyristor V1/load/thyristor V6/U_2.

After a phase shift of 60° (after V1 has fired) a pulse arrives at the gate of thyristor V2, the latter thereby becoming conducting. The current commutates from thyristor V6 to V2. The following circuit (time B) is produced:

U_1/thyristor V1/load/thyristor V2/U_3.

Thyristor V1 receives simultaneously as a second pulse the same trigger pulse which the thyristor V2 received as a main pulse (not shown in Fig. 7.64c; cf. Fig. 7.60). Thyristor V3 receives the main pulse shifted by a further 60° which also reaches thyristor V2 as a second pulse. Thyristor V3 thus becomes conducting and the current commutates from thyristor V1 to V3. The following circuit is now produced (time C):

U_2/thyristor V3/load/thyristor V2/U_3.

Figure 7.65 shows the behaviour of the rectified voltage at control angles α of 0°, 30°, 60° and 90°. As can be readily seen, with an increasing control angle the ideal no-load d.c. voltage $U_{di\alpha}$ becomes smaller. U_{12}, U_{13}, etc., designate the linked line voltages between the phases.

Figure 7.66 illustrates the control characteristic of the fully controlled three-phase bridge circuit.

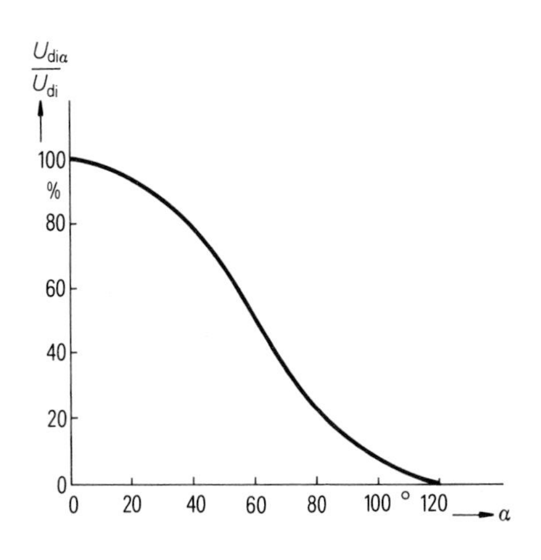

Fig. 7.66
Control characteristic of the fully controlled three-phase bridge circuit

Alternating current controller circuit

With alternating current controller circuits, alternating voltage can be modified infinitely and with little loss, i.e. adapted to the a.c. load (Fig. 7.67a). By varying the control angle α within the range from 0° to 180° the alternating voltage can be set between maximum and minimum (Fig. 7.67b). The back-to-back connected thyristor pair V1 and V2 can be fired by the trigger set at a certain time within the assigned semi-oscillation of the supply voltage. The respective conducting thyristor switches the voltage through to the load.

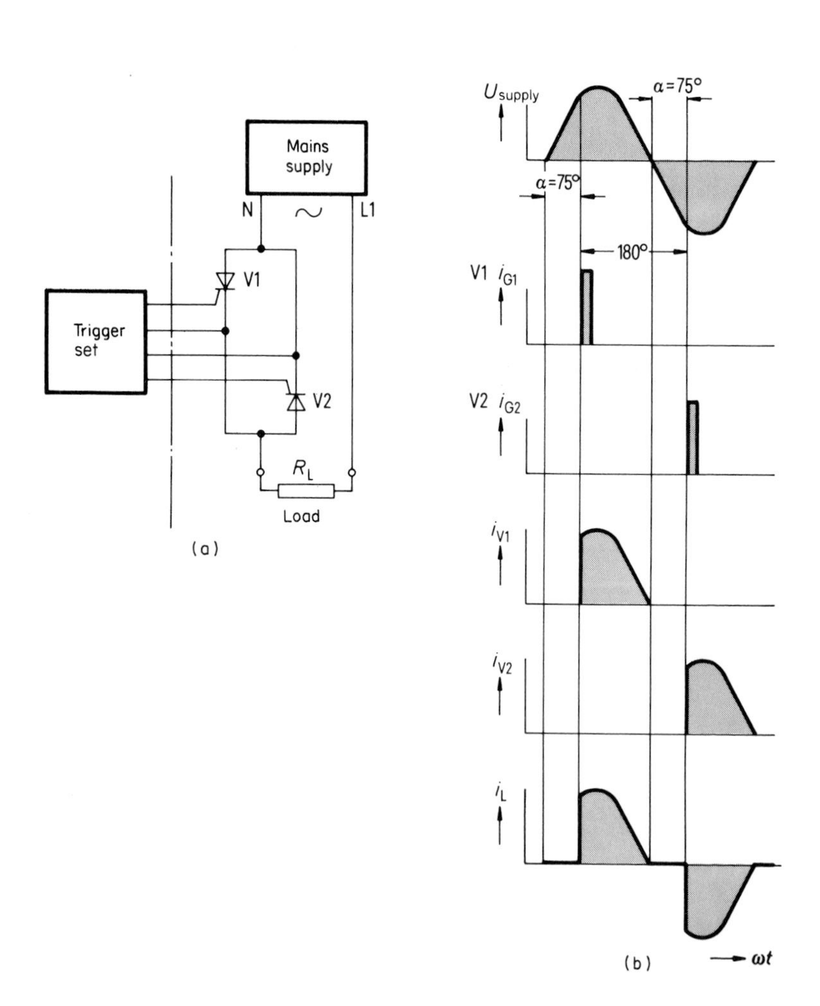

Fig. 7.67 Alternating current controller circuit

In contrast with the fully controlled three-phase bridge circuit (cf. Fig. 7.64), in the case of the fully controlled three-phase a.c. controller circuit the six thyristors are connected directly with the three-phase alternating supply voltage (Fig. 7.68). The arrangement is suitable for rectifiers with high powers (200, 500 and 1000 A GR 10 equipment).

As regards filtering requirements and loading of the mains with harmonics the circuit is comparable with the fully controlled three-phase bridge circuit.

The phase-shifting transformer additionally used with 500 and 1000 A rectifiers is explained in Section 5.3.1. This transformer is connected between the mains supply and the thyristor set A1.

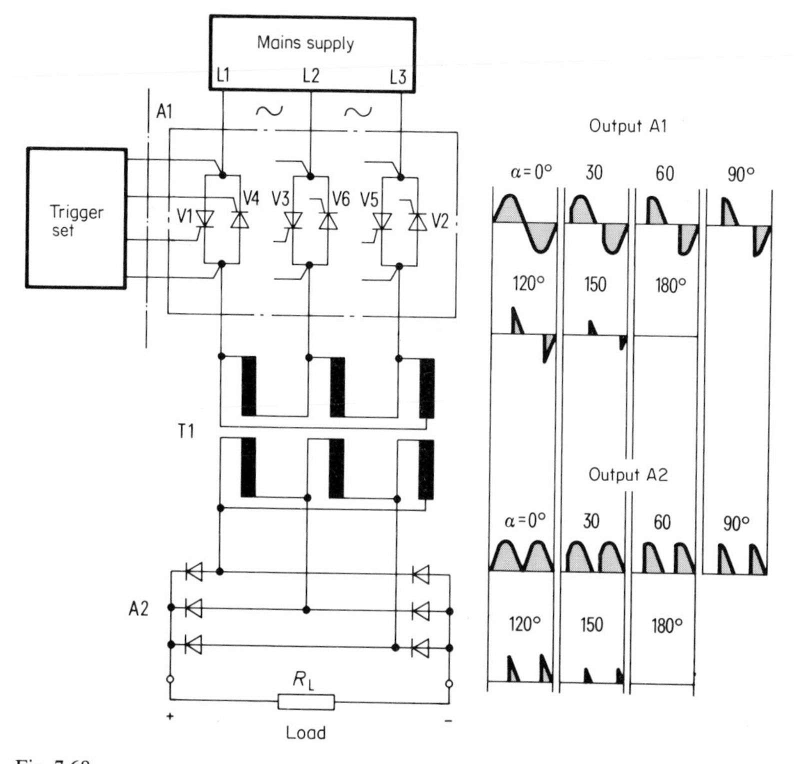

Fig. 7.68
Fully controlled three-phase a.c. controller circuit with single-phase representation of the output voltages of A1 and A2 as a function of the control angle α

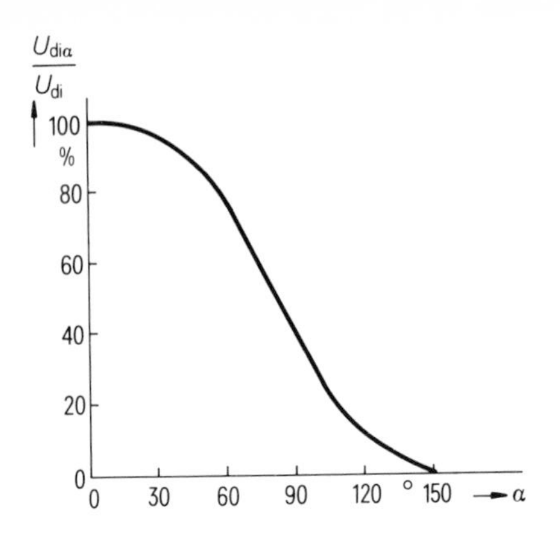

Fig. 7.69
Control characteristic of the fully controlled three-phase a.c. controller circuit

A relatively small current flows on the primary side of transformer T1 so that it is possible to use smaller (and therefore cheaper) thyristors. In this circuit the thyristor set (three-phase a.c. controller circuit A1) acts solely as a final control element. The trigger pulses are shifted in time so that there is a constant voltage at transformer T1 corresponding to the desired output voltage. The (three-phase) bridge circuit A2 takes over rectification of the voltage kept constant in controller A1.

Figure 7.69 shows the control characteristic for control angles from 0° to 150°.

Inverter circuit

The inverter circuit is used to convert a direct voltage, fed to it, for example, from a battery or rectifier, into alternating voltage.

Figure 7.70 shows an inverter in a centre circuit. The trigger set supplies the two thyristors V1 and V2 alternately with pulses. In this way current from the battery flows alternately through the transformer's two partial windings T1 and T2. A square-wave alternating voltage can be taken from the secondary side of the transformer, which for practical purposes is frequently converted into an approximately sinusoidal voltage by means of suitable arrangements. While it is well known that with rectifiers the thyristors are directly connected to alternating voltage and that extinguishing takes place at zero passage when the current drops below the hold current, in the case of an inverter a capacitor (C1) has to be provided for the extinguishing process.

If thyristor V1 receives a trigger pulse from the trigger set, its resistance becomes low.

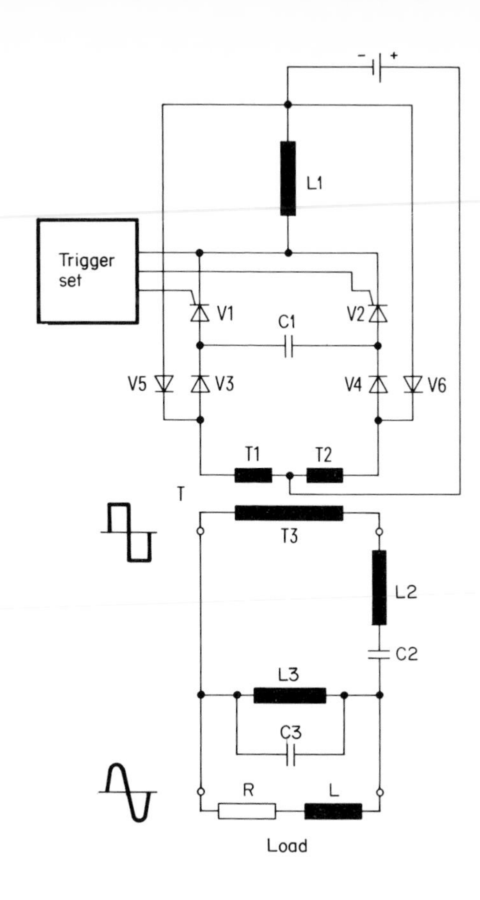

Fig. 7.70
Inverter in centre circuit

Circuit:

Battery positive terminal/partial winding T1/V3/V1/choke L1/battery negative terminal.

A voltage is induced in the secondary winding T3 and the capacitor C1 is charged via T2 and V4. The next trigger pulse goes to thyristor V2. Now *both* thyristors V1 and V2 are conducting; the capacitor discharges. The discharge current via V2 and V1 acts in thyristor V1 against the previous flow of current. V1 thereby becomes non-conducting, while V2 remains conducting.

Circuit:

Battery positive terminal/partial winding T2/V4/V2/L1/battery negative terminal.

A voltage is again induced in the secondary winding T3. The capacitor C1 is charged, though now via T1 and V3 with reversed polarity, etc.

Choke L1 limits the current on commutation. Diodes V3 and V4 prevent the quenching capacitor C1 from partially discharging prematurely via the primary windings of the transformer, while diodes V5 and V6 permit the connection of reactances. Part of the reactive current returns to the battery via these diodes.

Direct current controller circuit

With the aid of the direct current controller circuit (d.c. chopper controllers) direct voltage, e.g. battery voltage, can be converted into direct voltage at a different level (Fig. 7.71). A commutating device specifically turns off thyristor V1 at the respective desired moment.

If quenching thyristor V3 receives a trigger pulse, commutating capacitor C is charged.

Circuit:

> Battery positive terminal/capacitor C/quenching thyristor V3 (anode/cathode)/load/battery negative terminal.

The positive pole of the battery voltage is now on the left at capacitor C and the negative pole on the right.

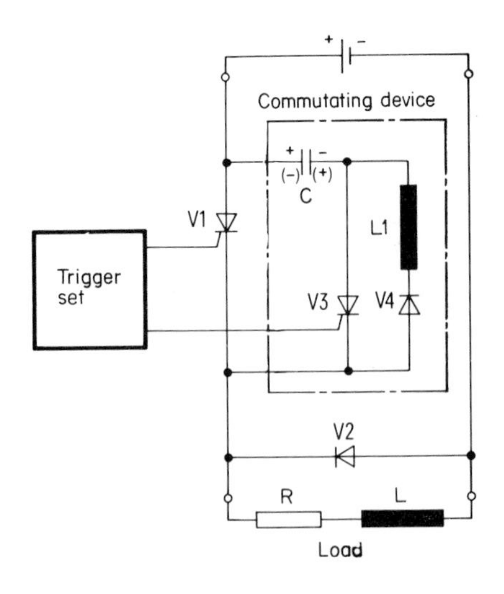

Fig. 7.71
Direct current controller circuit

The main thyristor V1 is then fired by the trigger set. The battery voltage is now applied to the load and a 'circuit in reverse' is formed, i.e. the capacitor C is inversely charged.

Circuit:

Capacitor C/main thyristor V1 (anode/cathode)/reverse diode V4/reverse choke L1/capacitor C.

The characters without brackets at the capacitor (plus/minus) apply to the time before reversing, those with brackets to the time after reversing.

The capacitor C is 'prepared' to quench the main thyristor V1 (when there is minus on the left and plus on the right) when the reversal process is ended.

Quenching thyristor V3 is then refired by the trigger set so that the capacitor voltage is in parallel with the main thyristor V1 (minus at anode and plus at cathode). Thyristor V1 is now in the off condition. The current continues flowing via the free-running diode V2. A mean value for the direct voltage $U_{di\alpha}$ is produced by the continuous alternate switching on and off of V1.

With d.c. chopper controllers use is made of:

pulse-width control and/or
pulse-sequence control (frequency control).

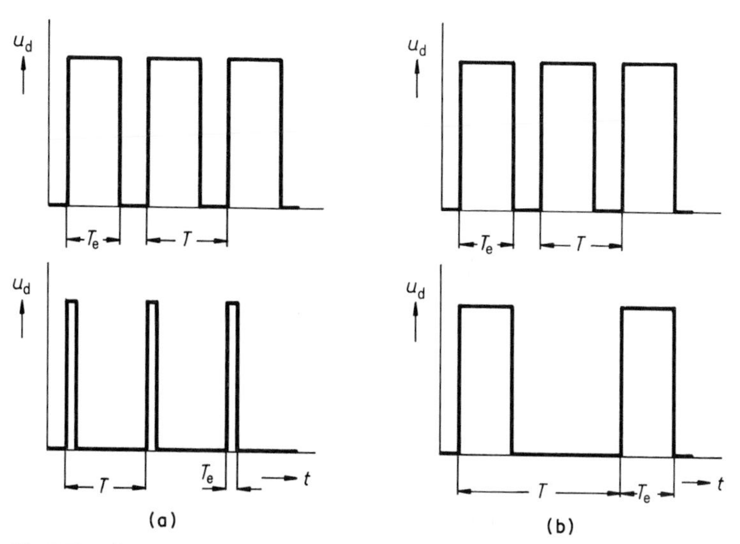

Fig. 7.72 a) Pulse-width control and b) pulse-sequence control

Pulse-width control, which is preferred in practice, involves the modification of the pulse-interval ratio by varying the on time T_e with the cycle duration T constant (Fig. 7.72a). A long on time results in a high output d.c. voltage and vice versa.

If the cycle duration T is changed (Fig. 7.72b), with a constant on time T_e, this is called the pulse-sequence control.

For the sake of completeness mention is also made in this connection of two-point current control when both the on time T_e and cycle duration T can be varied.

7.7 Integrated Circuits

With integrated circuits a distinction is made between:

monolithic circuits and
hybrid circuits.

Monolithic circuits (Fig. 7.73) are made using MOS and bipolar techniques. They permit the greatest possible integration density.

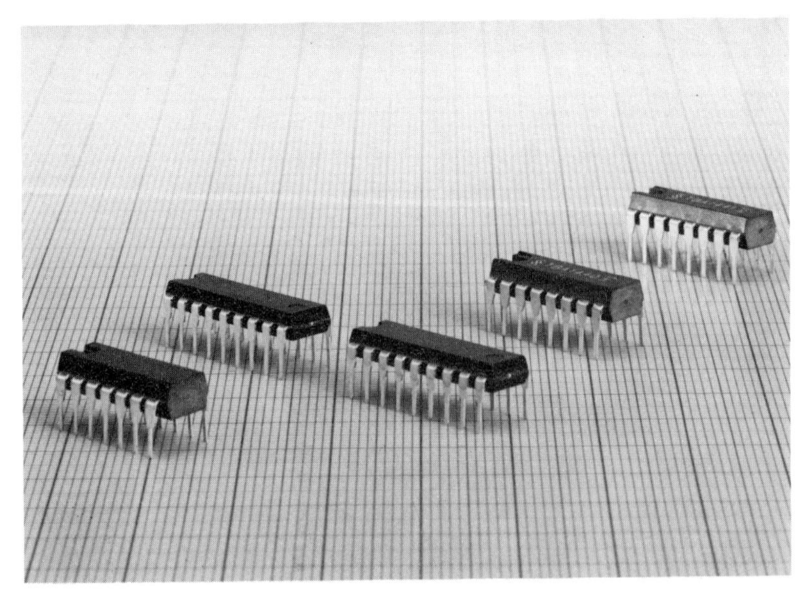

Fig. 7.73 Integrated circuits

Hybrid circuits are made up of discrete components (individual semiconductor devices) and integrated circuits. They contain on a mounting plate (glass, ceramic or plastic) passive components, e.g. coils, capacitors and resistors, and active components such as discrete semiconductor devices and integrated circuits.

Seen functionally, there are

analogue (integrated) circuits (e.g. operational amplifiers) and
digital (integrated) circuits.

The latter can be divided into

switching elements and
storage elements.

7.7.1 Operational amplifiers

The *ideal operational amplifier* can be described as follows:

- ▷ The voltage amplification V_u is very high; it designates the ratio of the change in output voltage ΔU_a to the change in input voltage ΔU_e.
- ▷ The input control power P_e is very low; it is the product of the change in input voltage ΔU_e and the change in input current ΔI_e.
- ▷ The transfer resistance R_t is very high: by this magnitude is understood the ratio of the change in output voltage ΔU_a to the change in input current ΔI_e.
- ▷ The output resistance R_a is very low; it designates the ratio of the change in output voltage ΔU_a to the change in output current ΔI_a.

According to the general representation of the amplifier (Fig. 7.74) the voltage at the input terminals is shown as U_e and the output voltage as U_a. This system of designations is also adopted for the operational amplifier (op amp), also termed a computing or variable-gain amplifier. The *graphical symbols* and *terminal markings* used in practice for the op amp are shown in Fig. 7.75.

In Fig. 7.75(a) can be seen the inverting input terminal E_-, the non-inverting input terminal E_+ and the output terminal A. Often in circuit diagrams the input terminals

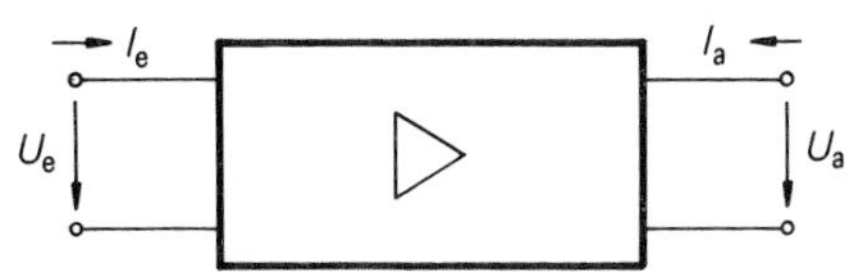

Fig. 7.74 General representation of the amplifier

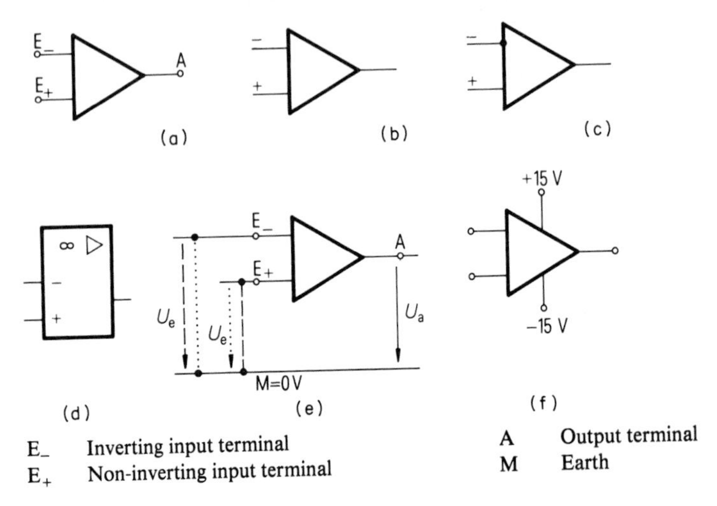

Fig. 7.75
Graphical symbols and terminal markings for the operational amplifier

are simply marked – and + (Fig. 7.75b). Sometimes, as practised in digital engineering, the inverting input terminal is shown with an inversion point (Fig. 7.75c). Occasionally the representation in Fig. 7.75(d) is found.

The voltages U_e and U_a are usually shown against a common reference potential M (earth) = 0 V (Fig. 7.75e). One of the two input terminals E_- or E_+ is frequently put to the reference potential across a resistance.

In Fig. 7.75(e), as an example, the input terminal E_+ is connected with M (broken line). The output voltage U_a is then reversed (inverted) in its polarity compared with the input voltage U_e (broken line) at the inverting input terminal E_- (in respect of M). If the input terminal E_- is connected with M (dotted line) and the voltage U_e (also dotted) is related to M, then the polarities of U_a and U_e are the same.

Sometimes the voltages at the input terminals are compared with each other; the following then applies. If the voltage at the inverting input terminal E_- is *positive* compared with the voltage at the non-inverting input terminal E_+, then there is voltage of negative polarity at the output terminal A and vice versa.

It can be seen from Fig. 7.76 that a small change in input voltage ΔU_e results in a large change in the output voltage ΔU_a. The 'steepness' of the *transfer characteristic* is a measure of the voltage amplification factor V_u. The saturation range is determined by the positive and negative saturation voltages. In the case of the 'overloaded' operational amplifier it depends on the supply voltage (operating voltage, e.g. 15 V; cf. Fig. 7.75f).

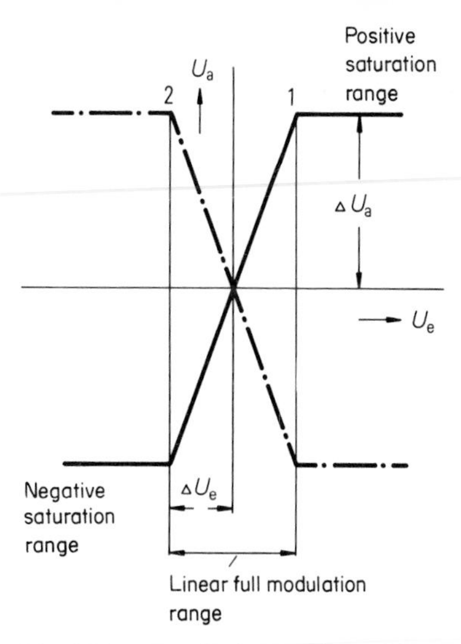

Fig. 7.76 Transfer characteristic of the operational amplifier

The transfer characteristic 1, drawn with a thick line, shows that if a voltage is applied to the non-inverting input terminal E_+, a non-inverted, positive polarity of the output voltage results. An increase in the value of the input voltage produces an increase in the voltage at the non-inverting terminal. The transfer characteristic 2 (dot–dash line) shows the inverted output voltage which depends on the negative voltage of the input terminal E_-. In the case of inversion, the output voltage drops when the input voltage rises.

Feedback circuit

The amplification characteristics can be modified by feeding back the output signal. Figure 7.77 shows the block diagram of a feedback circuit; in its simplest case it consists of a resistor R_f (cf. Figs. 7.78 and 7.79).

Feedback circuits can be designed as *negative feedback* or as *positive feedback.*

▷ *Negative feedback:* With negative feedback the part of the output signal conducted via the feedback circuit counteracts the input signal. This produces less amplification. The amplifier circuit is also stabilized. Normally the part of the output signal returned to the input is in anti-phase to the input signal (180° phase shift).

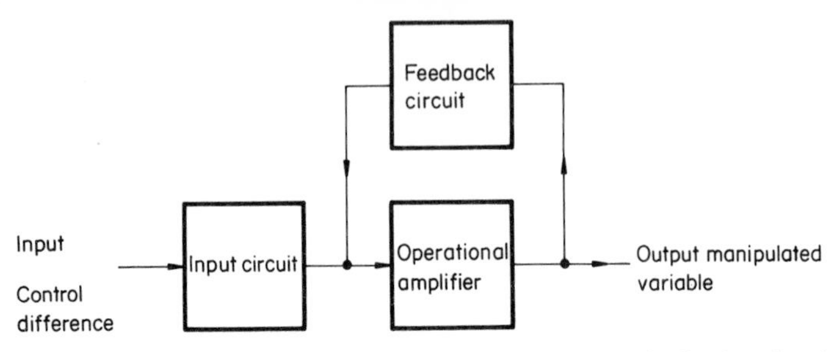

Fig. 7.77 Block diagram of an operational amplifier with a feedback and an input circuit

▷ *Positive feedback:* With positive feedback the effect of the output signal is to strengthen the input signal; in the extreme case this leads to oscillation of the circuit (circuits with astable flip-flop characteristic).

In the ideal case one would like with positive feedback to achieve zero phase shift between that part of the output signal returned to the input and the input signal.

Inverting amplifier with operational amplifier

The input terminal E_- can be wired via a number of inputs (broken line). Terminal E_+ is connected with the reference potential $M = 0$ V via the resistor R_M (Fig. 7.78). The input terminal E_- is used as the input voltage terminal. If the voltage at E_- is positive to M, then the output terminal A has negative polarity and vice versa (inversion).

The characters $-U_a$ indicate that the output is phase-shifted 180° compared with the inverting input E_-.

The input voltage U_0 causes a voltage drop at resistor R_0. The current I_0 flows. The current $-I_f$ returned via resistor R_f counteracts the input current (negative feedback).

The 'proportional amplification' V_R of the (ideal) inverting amplifier results from the ratio of the resistors in the input and feedback circuits.

The input resistance is determined above all by R_0; the output resistance is low. If the resistances R_f and R_0 are equally large, then the output voltage $-U_a$ is equal to the input voltage U_0. The inverting amplifier is used in a telecommunications power supply system as a controller. Often, for example, the set-point value is linked with the actual value at the inverting input and a controlled variable is formed (cf. Chapter 8).

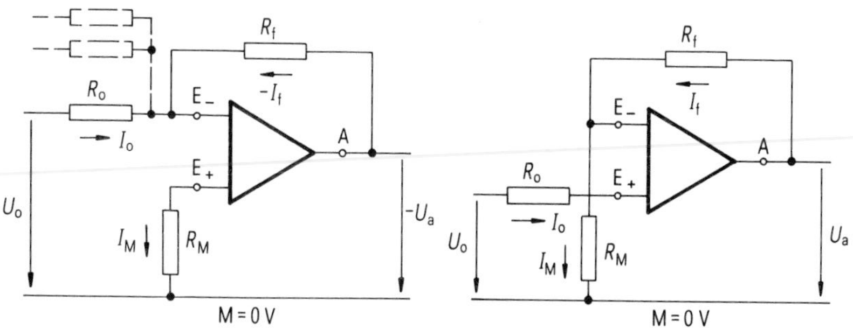

Fig. 7.78
Inverting amplifier with operational amplifier

Fig. 7.79
Non-inverting amplifier with operational amplifier

Non-inverting amplifier with operational amplifier

In Figure 7.79 the input terminal E_+ is used as the input voltage terminal. The input terminal E_- is connected with the reference potential $M = 0$ V via resistor R_M. There is no phase shift between the output voltage U_a and input voltage U_0. If the voltage at the input terminal E_+ is positive to M, then the voltage at the output terminal A is similarly of positive polarity and vice versa.

Proportional amplification V_R, as with the (ideal) inverting amplifier, depends on the ratio of the resistors in the input and feedback circuits. The input resistance with this circuit is very high.

The non-inverting amplifier is used, like the inverting amplifier, as a controller.

Matching transformer with operational amplifier

If, in the case of the non-inverting amplifier (cf. Fig. 7.79), the output terminal A is connected directly with the input terminal E_-, the output voltage U_a is as large as the input voltage U_0.

$V_R = 1$ applies to proportional amplification. This circuit achieves the highest possible input resistance and the lowest possible output resistance (Fig. 7.80).

The matching transformer is frequently used in telecommunications power supply equipment. It is used for decoupling function units and is also called the impedance matching transformer or impedance converter.

Differential amplifier with operational amplifier

Figure 7.81 shows the circuit of the differential amplifier. The output voltage $-U_a$

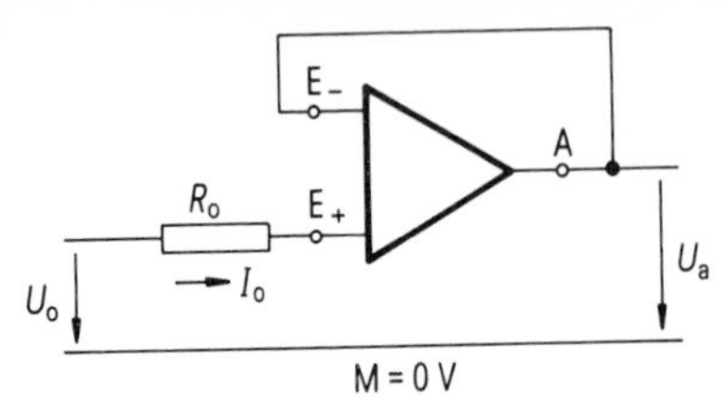

Fig. 7.80 Matching transformer with operational amplifier

depends on the difference between the two input voltages U_{01} and U_{02}. If in the circuit shown the input voltage U_{01} at the input terminal E_- is more positive than the input voltage U_{02} at the input terminal E_+, then the output voltage U_a at the output terminal is negative. If, on the other hand, the input voltage U_{01} is more negative than the input voltage U_{02}, the output voltage U_a becomes positive.

The differential amplifier is used, for example, in circuits for amplifying the current's actual value or for balancing the current (load-balancing) between a number of rectifiers.

Comparator with operational amplifier

The comparator (Fig. 7.82) compares the voltage U_{01} at the input terminal E_- with a reference voltage U_{02} at the input terminal E_+.

Even a few millivolts are sufficient to steer it into the positive or negative saturation range (cf. Fig. 7.76). A maximum positive output voltage U_a can be drawn at the

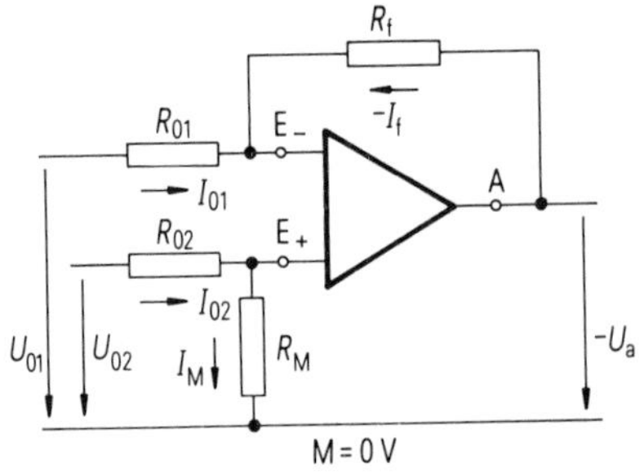

Fig. 7.81 Differential amplifier with operational amplifier

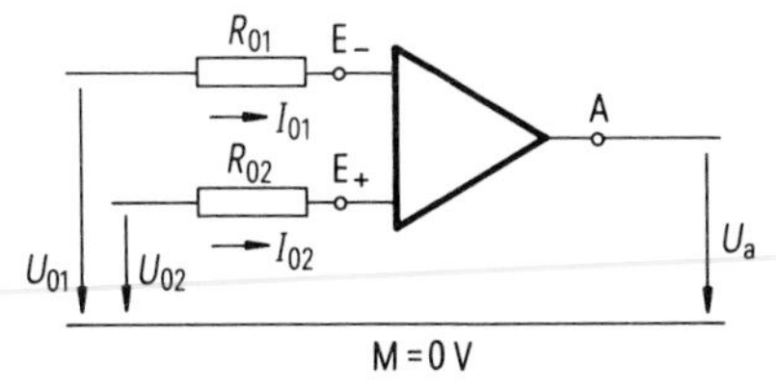

Fig. 7.82 Comparator with operational amplifier

output if the signal voltage U_{01} is more negative than the reference voltage. For a maximum negative output voltage U_{01} must be more positive than U_{02}.

The comparator is often used as a switch.

7.7.2 Switching elements

Digital switching elements (gates) are used for linking information (Fig. 7.83).

The three elementary switching elements are:

AND circuit,
OR circuit,
NOT circuit.

Among the 'extended' elements used in a telecommunications power supply system are:

NAND circuit,
NOR circuit,
EXCLUSIVE-OR circuit.

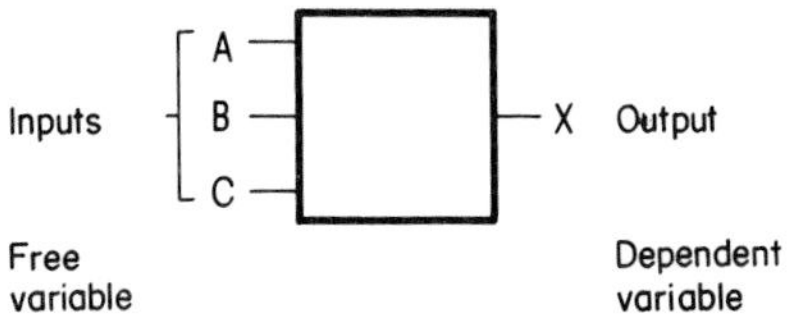

Fig. 7.83 Switching element with inputs A, B and C plus output X

In digital circuits the *signal statuses* are shown with L and H (Table 7.1). The H or L signal is often also called H or L level, or H or L for short.

Table 7.1 Signal statuses

0 L	1≙H
L (low)	**H (high)**
L, e.g. +4.5 V	H, e.g. +14 V

AND element

Figure 7.84 shows the example of an AND circuit using relay technology. If either only switch A (≙H signal) or only switch B (≙H signal) is closed, lamp X cannot light (≙L signal). Of course, it cannot also light if both switches are open. *Lamp X* can only *light* (≙H signal) if both switches A *and* B are *closed*. This can also be seen from the function table, also called a truth table (Fig. 7.85c).

The graphical symbols for the AND element as in DIN are shown in Fig. 7.85(a old) and (b new). From the truth table it can be established that:

> **The AND element always delivers an H signal to its output *X* when there is an H signal at *all* inputs.**

If only *one* input has an L signal, then there is also an L signal at the output. This can be expressed using Boolean algebra (Fig. 7.85d):

$$A \wedge B = X.$$

A pulse diagram (Fig. 7.86) is also used here for a better understanding of the working of the AND element.

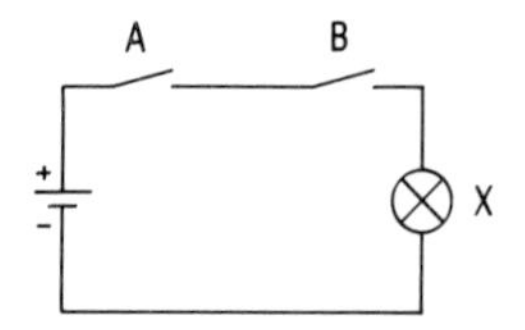

Fig. 7.84 AND circuit, showing switches

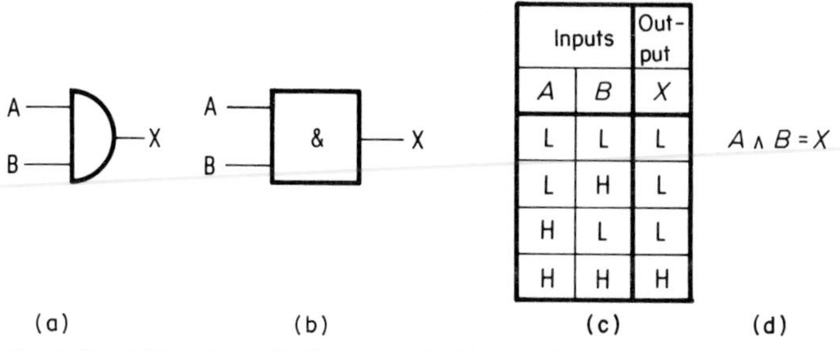

Inputs		Output
A	B	X
L	L	L
L	H	L
H	L	L
H	H	H

Fig. 7.85 AND circuit (AND element) with graphical symbols and truth table

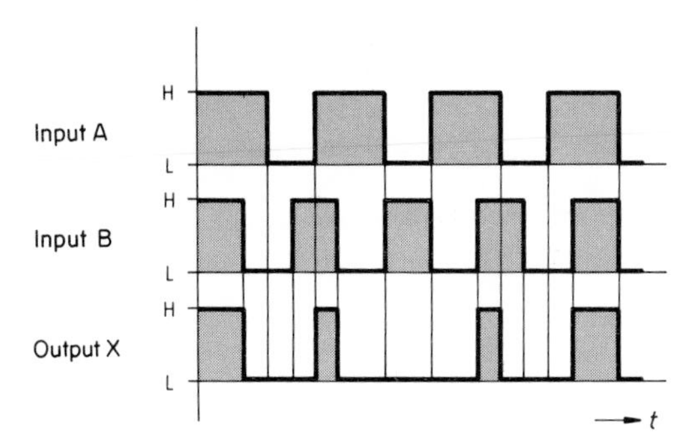

Fig. 7.86 Pulse diagram of the AND element

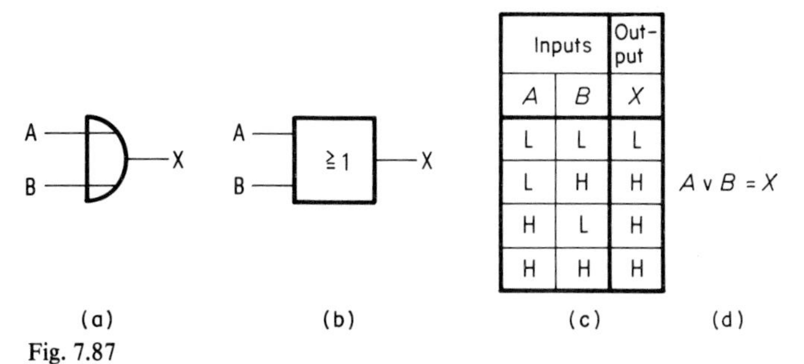

Inputs		Output
A	B	X
L	L	L
L	H	H
H	L	H
H	H	H

Fig. 7.87
OR circuit (OR element) with graphical symbols and truth table

OR element

Figure 7.87 shows the graphical symbols, function table and Boolean expression for the OR element. Thus:

The OR element delivers an H signal to its output *X* if there is an H signal at *one* input at least.

As can be seen from the truth table, there is only an L signal at the output when both inputs have an L signal.

NOT element

The output of the NOT element is always the opposite signal to that at the input (Fig. 7.88).

In the graphical symbol in Fig. 7.88(a) a *NOT point* (inversion point) can be seen (top graphical symbol); this indicates the signal reversal. In the graphical symbol in Fig. 7.88(b) the inversion is marked by a *NOT circle* at the output (top graphical symbol). The NOT circle or NOT point can also be shown at the input of the graphical symbol (Fig. 7.88b and a, bottom graphical symbols).

It follows from the function table (Fig. 7.88c) that:

With the NOT element there is always at the output *X* the *reversed* signal of the input *A*.

The Boolean expression is (Fig. 7.88d):

$$A = \overline{X}.$$

NAND element

The NAND element represents a combination of an AND element with a NOT one. Figure 7.89 shows an AND element with the inputs A and B and the output X. The output X becomes input A for the following NOT circuit.

When both successively connected graphical symbols, as shown in Fig. 7.89, are 'combined' into one graphical symbol, one obtains the graphical symbol for the NAND circuit (Fig. 7.90a and b, top graphical symbols). The bottom graphical symbols also represent NAND elements (input-negated OR circuit).

It can be seen from the function table (Fig. 7.90c) that:

The NAND element always supplies an L signal to its output *X* only when there is an H signal at *all* inputs.

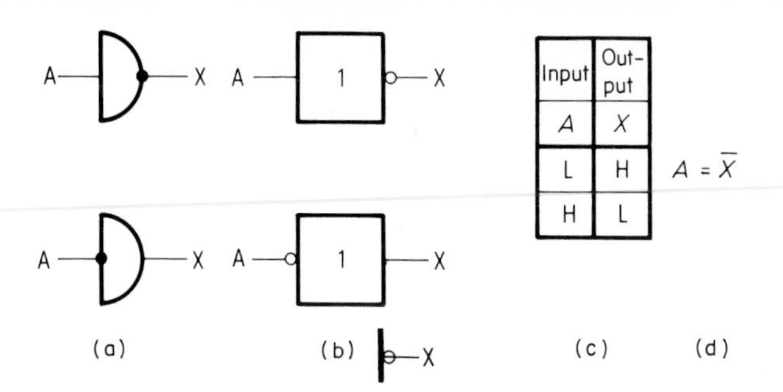

Input	Output
A	X
L	H
H	L

Fig. 7.88
NOT circuit (inverter, NOT element) with graphical symbols and truth table

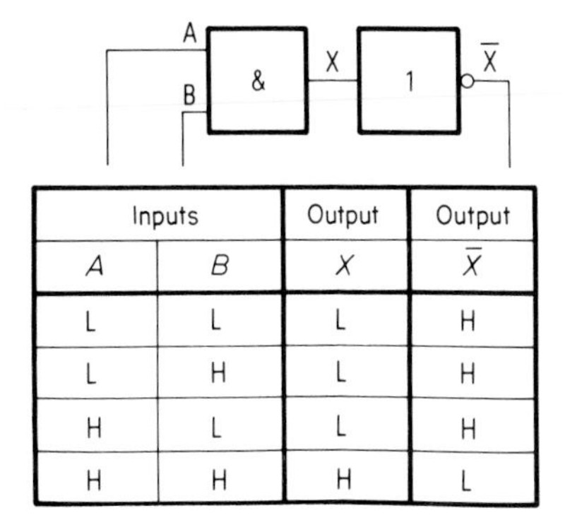

Inputs		Output	Output
A	B	X	$\overline{X}$
L	L	L	H
L	H	L	H
H	L	L	H
H	H	H	L

Fig. 7.89 AND circuit with following NOT circuit

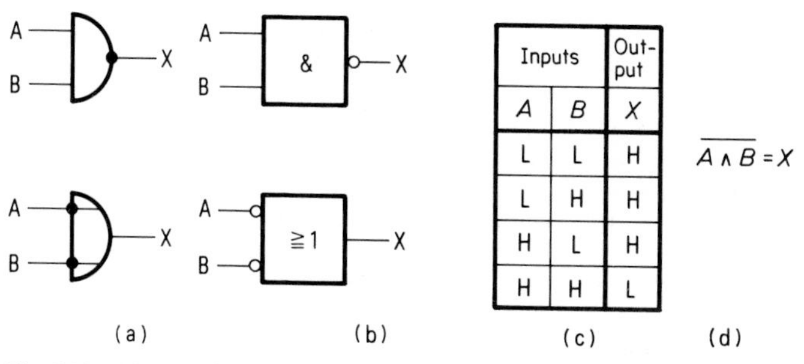

Inputs		Output
A	B	X
L	L	H
L	H	H
H	L	H
H	H	L

Fig. 7.90 NAND circuit with graphical symbols and truth table

An L signal at one of the inputs forces the output to an H signal. This is also shown by the Boolean expression (Fig. 7.90d).:

$$\overline{A \wedge B} = X.$$

NOR element

The NOR element is produced by combining an OR element with a NOT one (Fig. 7.91a and b, top graphical symbols). The bottom graphical symbols also represent NOR elements (input-negated AND circuits).

It can be seen from the function table (Fig. 7.91c) that:

The NOR element always supplies an L signal to its output *X* when there is an H signal at *one* input at least.

The Boolean expression (Fig. 7.91d) is:

$$\overline{A \vee B} = X.$$

EXCLUSIVE-OR element

There is only an H signal at the output *X* if the input signals are *not the same*; thus only one input is H and all other inputs are L. If the inputs are *the same*, then there is always an L at the output (Fig. 7.92). The EXCLUSIVE-OR element is marked either with a small e in the OR graphical symbol (Fig. 7.92a) or with =1 (Fig. 7.92b).

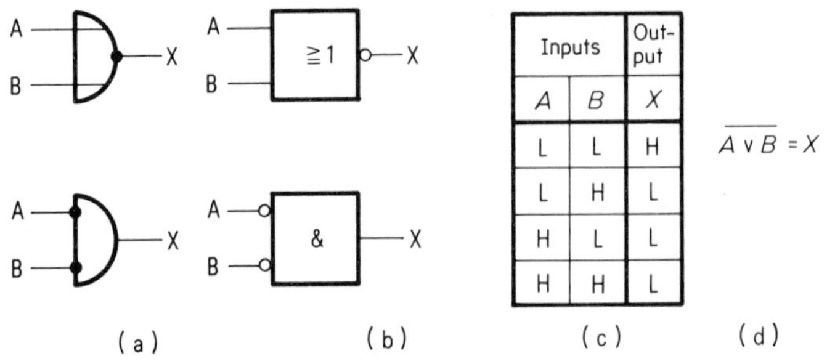

Inputs		Output
A	*B*	*X*
L	L	H
L	H	L
H	L	L
H	H	L

Fig. 7.91 NOR circuit with graphical symbols and truth table

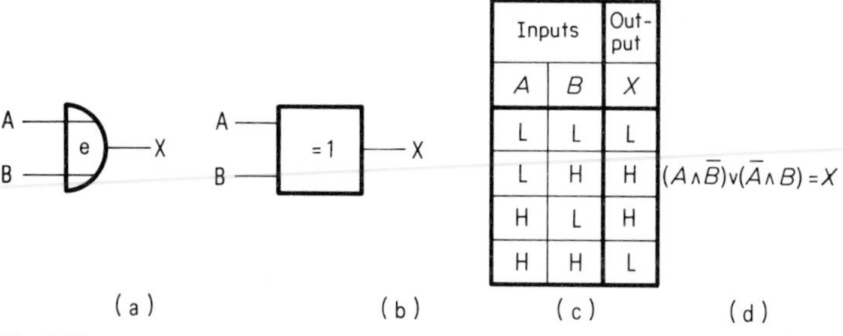

Inputs		Output
A	B	X
L	L	L
L	H	H
H	L	H
H	H	L

Fig. 7.92
EXCLUSIVE-OR circuit (EXOR, anti-coincidence element) with graphical symbols and truth table

It can be seen from the function table (Fig. 7.92c) that:

The EXCLUSIVE-OR element supplies an H signal to its output *X* only when there is an H signal at only *one* input.

The Boolean equation (Fig. 7.92d) is:

$$(A \wedge \overline{B}) \vee (\overline{A} \wedge B) = X.$$

7.7.3 Storage elements

The task of a storage element is to store binary signals with or without a time limit. Simple storage elements are made with trigger circuits.

A distinction is made between

bistable toggle stages (flip-flop),
monostable toggle stages (monoflop) and
astable toggle stages.

Bistable toggle stages represent storage elements without a time limit and are stable in the set and reset state; they cannot *themselves* alter the respective state.

By *monostable toggle stages* are meant time-limited storage elements; they are stable in *one* state only (usually in the reset state). When a monostable toggle stage is set, it only remains a certain time in this state and then *of its own accord* flips into the other state.

Astable toggle stages oscillate continuously to and fro between two states.

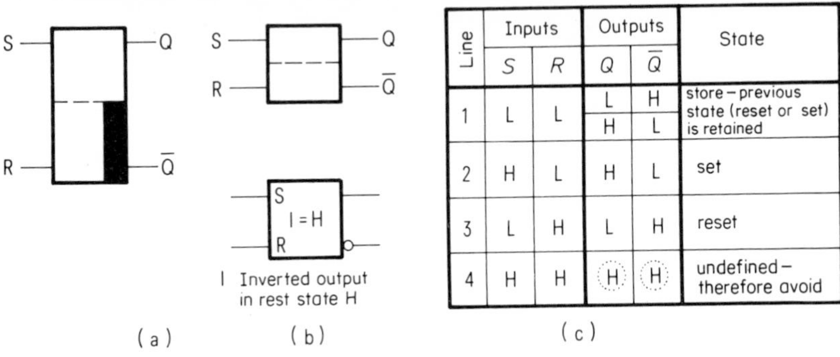

Line	Inputs		Outputs		State
	S	R	Q	$\bar{Q}$	
1	L	L	L / H	H / L	store – previous state (reset or set) is retained
2	H	L	H	L	set
3	L	H	L	H	reset
4	H	H	H	H	undefined – therefore avoid

Fig. 7.93 RS flip-flop with graphical symbols and truth table

Bistable toggle stages

RS flip-flop (= *reset/set*): The simplest bistable toggle stage is the RS flip-flop (Fig. 7.93). In the graphical symbols (Fig. 7.93a and b) can be seen at the top the set input S and the set output Q. Below is the reset input R and reset output $\bar{Q}$.

After applying the operating voltage the RS flip-flop adopts a basic position (rest position). If the small field on the right of the reset field is filled in with black (Fig. 7.93a), this indicates that the RS flip-flop is reset in the rest state. An RS flip-flop set in the rest position is shown by an unfilled-in field in the reset part. As in Fig. 7.93(b) the broken line between the top set field and the bottom reset field can be omitted. It is also permissible to show the reset output $\bar{Q}$ by means of a NOT circle.

The function table (Fig. 7.93c) shows that the outputs remain unchanged (first line) if both inputs have an L signal. As the state that was prevailing at the time is retained, it is called 'storing'. A change of signal from L to H at the set input S results in an H signal at output Q and an L signal at output $\bar{Q}$ ('setting' the store, second line).

If it is required to reset the store, a change of signal from L to H must be made at the reset input R. There is then an L signal at output Q and an H signal at output $\bar{Q}$ (third line). If both inputs simultaneously receive a change in signal from L to H, then the RS flip-flop passes into an undefined state (fourth line).

A pulse diagram (Fig. 7.94) is also given here for a better understanding of the function of the RS flip-flop.

The RS flip-flop is set by a change of signal from L to H at set input S (time t_1). The set output Q now changes from L to H. At the reset output $\bar{Q}$ the signal changes from H to L.

If the state changes from L to H at the reset input R, the RS flip-flop is reset (time t_2). The set output Q now flops back to L and the reset output $\bar{Q}$ again to H until the process is repeated with the next pulse at the set input S. At time t_3 the RS flip-flop is again set and at time t_4 again reset.

In both stable positions (set and reset state) the two outputs Q and $\bar{Q}$ always have opposite signals (cf. Fig. 7.93c). If an already set RS flip-flop receives a second set pulse, the states of the outputs do not change; even a second reset pulse can no longer influence the RS flip-flop.

An RS flip-flop can also be made from two NAND circuit elements or with two NOR circuit elements (Fig. 7.95).

JK flip-flop: In contrast with the RS flip-flop, with which the set input and reset input are static state-controlled inputs (Fig. 7.96a), dynamic cycle-edge controlled inputs (e.g. as with the JK flip-flop) react to the active cycle edges (Fig. 7.96b and c).

The cycle input markings in Fig. 7.96(b) mean that the output signal is triggered with the leading edge (signal change from L to H). If the active cycle edge of a dynamic input is the trailing edge, the output signal is triggered on passing from H to L (Fig. 7.96c). The input T or C is called the trigger input (clock input).

Figure 7.97 shows the graphical symbols and function table of the JK flip-flop;

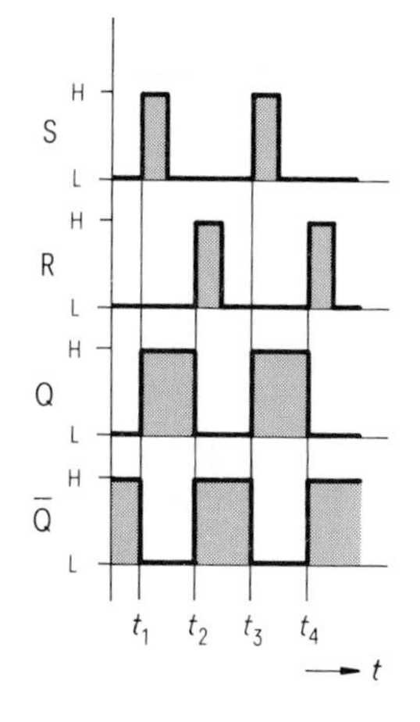

S Set input
R Reset input
Q Set output
$\bar{Q}$ Reset output

Fig. 7.94
Pulse diagram for RS flip-flop

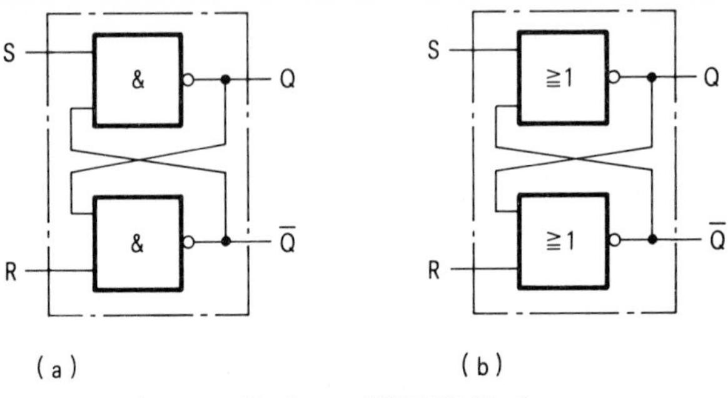

Fig. 7.95 a) NAND flip-flop and b) NOR flip-flop

apart from the dynamic cycle input T (or C) it also has two preparatory inputs J and K (Fig. 7.97a and b) which must become active *before* the cycle. Preparatory input J represents the set input and preparatory input K the reset one. There can also be preferential controlling inputs $\bar{S}$ and $\bar{R}$ present. An L signal at $\bar{S}$ causes setting of the JK flip-flop and an L signal at $\bar{R}$ its resetting, regardless of the state of the cycle input T and the state of the preparatory inputs J and K.

When both controlling inputs $\bar{S}$ and $\bar{R}$ have an L signal, there is an H signal at both outputs Q and $\bar{Q}$ (because, of course, a flip-flop cannot be simultaneously set and reset). If set input J and reset input K are used, both controlling inputs $\bar{S}$ and $\bar{R}$, must be on H signal, hence be inactive. Both controlling inputs $\bar{S}$ and $\bar{R}$ are the same in function as with the RS flip-flop, though with the JK flip-flop the active signal is usually L.

The JK flip-flop remains in its previous position (stored state) when both static preparatory inputs J and K have an L signal and then the clock pulse comes (first line

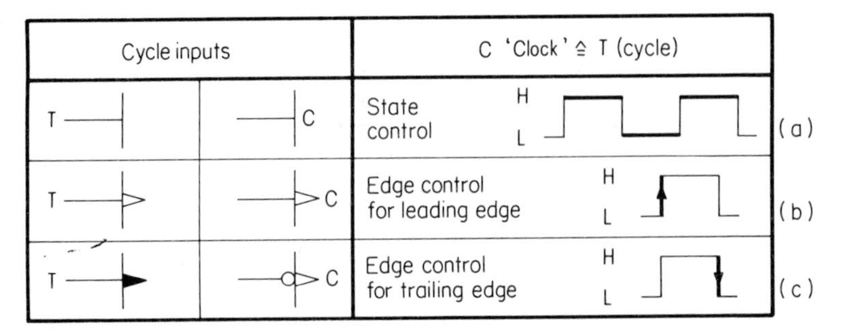

Fig. 7.96 Markings of cycle inputs

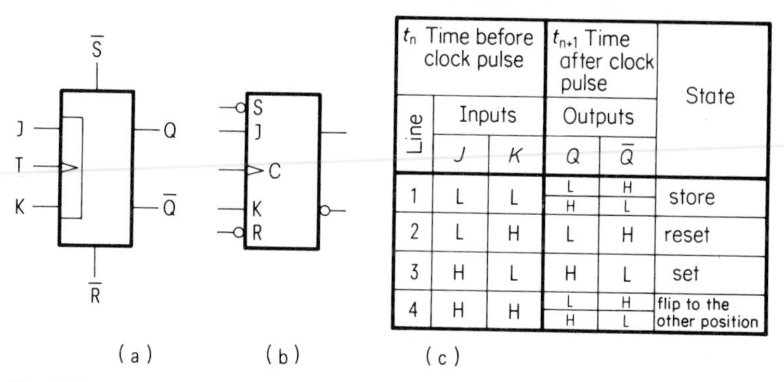

Line	t_n Time before clock pulse		t_{n+1} Time after clock pulse		State
	Inputs		Outputs		
	J	K	Q	$\bar{Q}$	
1	L	L	L / H	H / L	store
2	L	H	L	H	reset
3	H	L	H	L	set
4	H	H	L / H	H / L	flip to the other position

Fig. 7.97 JK flip-flop with graphical symbols and truth table

of function table, Fig. 7.97c). If there is an H signal at the preparatory input K before the clock pulse appears, the JK flip-flop is reset (second line). If there is an H signal at the preparatory input J before the clock pulse appears, the JK flip-flop is set (third line). If there is an H signal at both preparatory inputs J and K before the clock pulse arrives, the JK flip-flop is flipped into the other position. If it is set, it is reset and vice versa (fourth line).

JK master–slave flip-flop: The master–slave flip-flop can be made up of two JK flip-flops with common edge control. The master flip-flop determines which state the (following) slave flip-flop must adopt. The cycle inputs of both flip-flops are connected. The information is first accepted into the master flip-flop with the leading cycle edge (transition from L to H). The information is only stored there. With the following trailing edge, transition from H to L, the information passes on from master to slave. The information now also appears at the outputs (Fig. 7.98).

The triangle, entered in black at the cycle input T in Fig. 7.98(a), and the input-negated cycle input C with triangle in Fig. 7.98(b) indicate that the information of the J and K inputs appears at the outputs with the trailing edge of the timing signal. The character '⅂' means retarded (delayed) output and relates to the cycle input. The states at the outputs Q and Q̄ only change when the signal passes from H to L. With the newer graphical symbol (Fig. 7.98b) the master–slave flip-flop is only distinguished from the JK flip-flop (cf. Fig. 7.97b) by the retarded outputs.

The effect of the controlling inputs R̄ and S̄ is exactly the same as already described with the JK flip-flop.

The tables in Fig. 7.98(c) and (d) can provide a better understanding of the functioning of the master–slave flip-flop. According to Fig. 7.98(c) the controlling inputs S̄

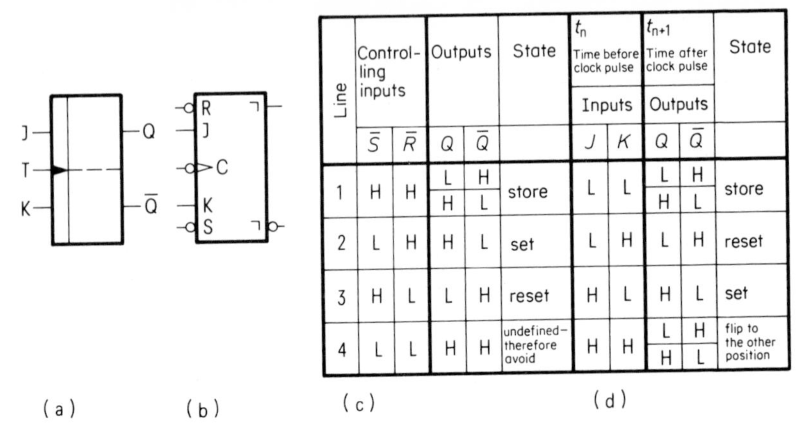

Line	Controlling inputs		Outputs		State	t_n Time before clock pulse: Inputs		t_{n+1} Time after clock pulse: Outputs		State
	$\bar{S}$	$\bar{R}$	Q	$\bar{Q}$		J	K	Q	$\bar{Q}$	
1	H	H	L / H	H / L	store	L	L	L / H	H / L	store
2	L	H	H	L	set	L	H	L	H	reset
3	H	L	L	H	reset	H	L	H	L	set
4	L	L	H	H	undefined– therefore avoid	H	H	L / H	H / L	flip to the other position

Fig. 7.98
JK master–slave flip-flop with graphical symbols and truth tables

and $\bar{R}$ work independently of the clock pulse. As with the JK flip-flop, setting is with an L signal at input $\bar{S}$ and resetting with an L signal at input $\bar{R}$.

In the representation in Fig. 7.98(d) the inputs J and K are dependent of the clock pulse. The information is accepted at the outputs with the trailing edge, when the clock pulse changes from H to L. The function table for the master–slave flip-flop (Fig. 7.98d) is the same as the function table for the JK flip-flop (cf. Fig. 7.97c).

Figure 7.99 shows the four phases of the action of the clock pulse.

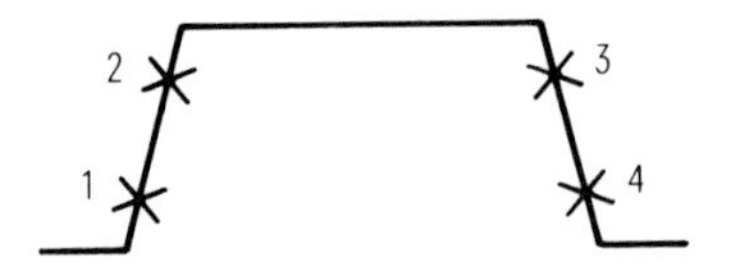

1 Separate slave from master
2 Input signal from J and K into master
3 Block J and K inputs
4 Transfer information from master to slave and hence to the outputs

Fig. 7.99 Clock pulse for master–slave flip-flop

Monostable toggle stages

The monostable toggle stage with a trigger pulse (clock pulse) can be temporarily brought from the stable to unstable state. After a delay time (dwell time) this toggle stage returns to its original position.

A distinction is made between non-retriggerable monostable toggle stages and retriggerable monostable toggle stages.

If a monostable toggle stage is connected to be 'non-retriggerable', further trigger pulses remain ineffective if it has already flipped into the unstable state. These cannot be effective until it has flopped back into the stable state after expiry of the delay time.

If a monostable toggle stage is connected to be 'retriggerable' (e.g. as a pulse delay), it can be retriggered again at any time during the unstable state, and thus during the delay time. With this the delay time restarts anew.

The graphical symbols for the monostable toggle stage are shown in Fig. 7.100. The dynamic trigger input T can be seen in Fig. 7.100(a). As the triangle is not filled in, the function of the trigger input is triggered with the leading edge of the trigger pulse (signal change from L to H). The arrow points to the field whose output signal is active in the stable position.

In this example the arrow is pointing down into the reset field. This means that in the rest position the reset output $\bar{Q}$ has an H signal and the set output Q has an L signal. The stable state here is therefore the reset state. Monostable toggle stages can have an inverted reset input $\bar{R}$. While the delay time is running resetting can be brought about by means of an L pulse to this input.

The cycle input (C) can be seen in the graphical symbol (Fig. 7.100b). In addition, in the newer symbol the monostable toggle stage is indicated by '1⎍'. This does not say anything though about the type and polarity of the output pulse.

The output marked with the NOT circle corresponds with the reset output shown with $\bar{Q}$ in the graphical symbol (Fig. 7.100a).

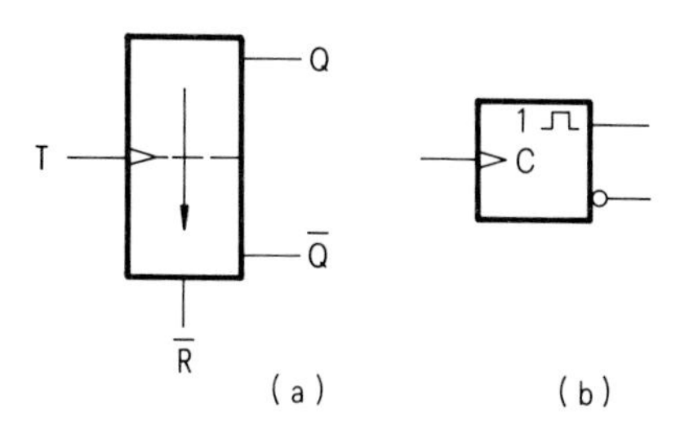

Fig. 7.100
Monostable toggle stage, graphical symbols

The difference between a *non-retriggerable* and a *retriggerable* monostable toggle stage is explained with the aid of pulse diagrams (Fig. 7.101).

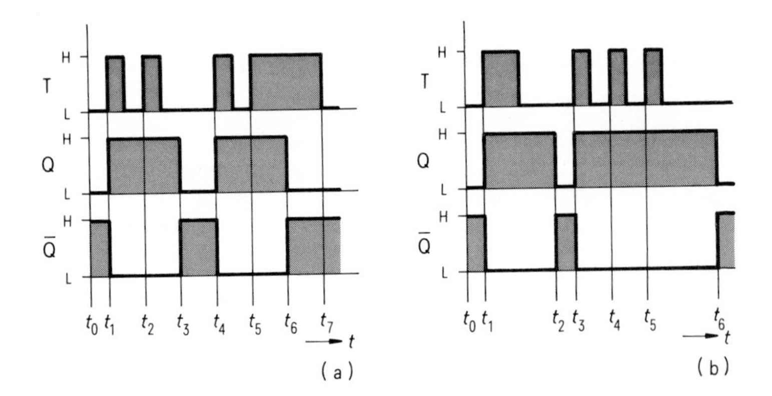

Fig. 7.101
Pulse diagram for a) non-retriggerable and b) retriggerable monostable toggle stages

Explanation of Fig. 7.101a

t_0	Rest position The dynamic trigger input T is at L The monoflop is reset to stable position $Q = L \quad \bar{Q} = H$
t_1	With the leading edge (L to H) at the dynamic trigger input T the monoflop is set to the unstable position. Both outputs thereby flip to the other position $Q = H \quad \bar{Q} = L$ The delay time starts to run (time t_1 to t_3)
t_2	A second trigger pulse in fact comes during the delay time of the monoflop. But despite the edge change from L to H of the trigger pulse the position of the monoflop does not change; it remains in set state. The delay time continues running
t_3	The delay time has now expired. The monoflop therefore of its own accord flops back automatically to the stable (reset) position $Q = L \quad \bar{Q} = H$
$t_4 = t_1$	
$t_5 = t_2$	
$t_6 = t_3$	
t_7	Although the last trigger pulse was longer than all others (time t_5 to time t_7), the monoflop still remains unaffected

Explanation of Fig. 7.101b

t_0	Rest position The dynamic trigger input T is at L The monoflop is reset to stable position $Q = L \quad \bar{Q} = H$
t_1	With the leading edge (L to H) at the dynamic trigger input T the monoflop is set to the unstable position. Both outputs thereby flip to the other position $Q = H \quad \bar{Q} = L$ The delay time starts to run (time t_1 to t_2)
t_2	The delay time has now expired. The monoflop therefore automatically flops back of its own accord to the stable (reset) position $Q = L \quad \bar{Q} = H$
$t_3 = t_1$	
t_4	Before the delay time has completely expired, the monoflop is retriggered with the next edge change from L to H at the dynamic trigger input T. The delay time thereby restarts afresh. There is therefore no change at the outputs Q and $\bar{Q}$ Q remains H $\quad \bar{Q}$ remains L The monoflop thus remains in set state
$t_5 = t_4$	
t_6	The delay time has now expired. The delay time range from t_5 to t_6 is the same as the delay time range from t_1 to t_2. As there was no retriggering within the delay time from t_5 to t_6, the monoflop at time t_6 automatically flops back of its own accord to the stable (reset) position $Q = L \quad \bar{Q} = H$

8 Application of Control Engineering in Power Supply Equipment

This chapter discusses in more detail the application of control engineering in modern power supply equipment. Some elementary correlations and terms in open-loop control and closed-loop control are presented for those readers for whom this part of the subject is not so familiar.

With rectifiers a distinction is made between uncontrolled, phase-controlled and controlled equipment.

With *uncontrolled rectifiers* the d.c. output voltage falls when the load increases. It is also dependent on fluctuations in the alternating supply voltage and supply frequency.

Phase-controlled rectifiers supply a constant d.c. output voltage with changes in load and fluctuations in the a.c. supply voltage; however, the d.c. output voltage does change in the case of fluctuations in the supply frequency.

Controlled rectifiers supply a constant d.c. voltage within the tolerance band of, for example, $\pm 0.5\%$, regardless of any change in load or fluctuations in supply voltage and frequency. They receive their energy either directly from the mains supply or from standby power supply systems and must meet both the requirements of the communications systems as well as those of the battery.

This group of rectifiers is classified into equipment with a transductor power section (magnetically controlled rectifiers), thyristor power section (thyristor-controlled rectifiers), and transistor power section (transistor-controlled rectifiers). With thyristor-controlled rectifiers there are those with phase-angle control and those with switching controller. Rectifiers with a transistor power section can be further divided into those with a longitudinal controller and those with a switching controller and switching-mode power supplies.

Another important group comprises *D.C./D.C. converters*; these are supplied with direct current and can form independent subassemblies and equipment and also constitute parts of switching-mode power supplies. Inverters should also be mentioned at this point, though they are not considered here (cf. Section 7.6.4).

8.1 Working

First are compared the ways in which open-loop and closed-loop control systems

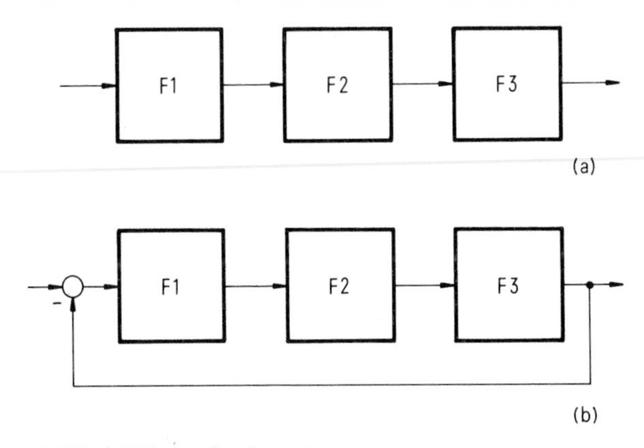

F1, F2 and F3 transfer elements

Fig. 8.1 Operating modes of a) open-loop control and b) closed-loop control

work (Fig. 8.1). The feature of open-loop control is the open control chain – that of closed-loop control of the closed-loop circuit.

In the case of *open-loop control* one or more imput quantities in a demarcated system influence the output quantity. The output quantity does not react upon the input quantity. A control system consists of the control equipment and the controlled system (Fig. 8.2a).

The command variable w arrives at the control equipment from an outside source. Depending on this command variable the control equipment generates a correcting variable y which exerts a controlling influence upon the controlled system. The controlled system represents the part of a system, the influencing of which is the subject in hand.

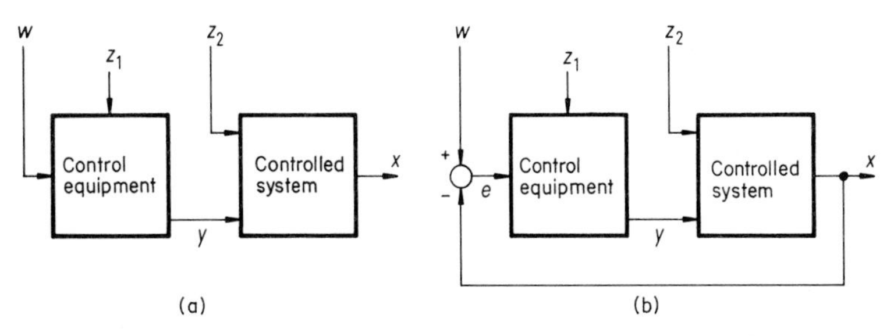

Fig. 8.2 Basic working of a) open-loop control and b) closed-loop control

The controlled object is also discussed. In this is formed the quantity to be controlled, the controlled variable. Disturbances z (e.g. supply voltage fluctuations or changes in load) interfere with the control equipment (disturbance z_1) and also with the controlled system (disturbance z_2) and impair the working of the open-loop control system. Disturbances acting from outside can result in the controlled variable varying considerably from the command variable. If the command variable w changes, this causes a change in the correcting variable y and thereby also in the output variable for the controlled system (control variable x).

Any disturbance thus influences the output variable. In the open-loop control system therefore the output variable is very restricted in how it can follow the command variable. If there is a constant disturbance, this can be compensated for by presetting the command variable accordingly. However, if the disturbances are constantly changing it is necessary, because of the disadvantages of the open-loop control system described, to use a closed-loop control system instead.

Closed-loop control can, despite disturbing influences, adjust as near as possible the controlled variable to the value preset by the command variable. A closed-loop control system consists of control equipment and a controlled system.

With closed-loop control (Fig. 8.2b) the variable to be controlled (controlled variable, actual value) is if necessary continuously logged, compared with the constant command variable (set-point value), and depending on the result of this comparison is influenced towards adjusting to the command variable. The result is an operation within a closed circuit, called the closed-loop control circuit. A fixed set command variable w (set point value) is compared at the input of the control equipment with the feedback controlled variable x (actual value). The difference between the command variable w and the controlled variable x is the control difference e. The aim of feedback control processes is always to keep the control difference between the command variable and controlled variable as small as possible. As long as there is a difference between the command variable and controlled variable, the control equipment reduces the difference between them by providing an appropriate correcting variable.

As with the open-loop control system, disturbances z_1 and z_2 can also affect the closed-loop control circuit, though here these effects, e.g. on the d.c. output voltage, can be corrected.

The closed-loop control system reacts to changes in the command variable or disturbances with a control process, i.e. the controller produces a correcting variable, which brings the controlled variable to the new set point value, respectively holds it at the preset set point. This control process is always associated with a transient. This means that the controller can only react when a control difference has occurred. Only then is the controlled variable brought either by oscillation or aperiodically to the new respectively original set point value.

8.2 Units of the Closed-Loop Control Circuit

Figure 8.3 shows the units making up the closed-loop control circuit.

8.2.1 Final control element

The flow of energy is present with uncontrolled voltage at the input of the final control element. In the case of rectifiers, for example, it is the thyristor set made up of thyristors together with its trigger set or the power transistor unit with its driving transistors. The final control element has the task, according to the correcting variable supplied by the control equipment, of controlling the flow of energy so that it is adapted to the controlled object.

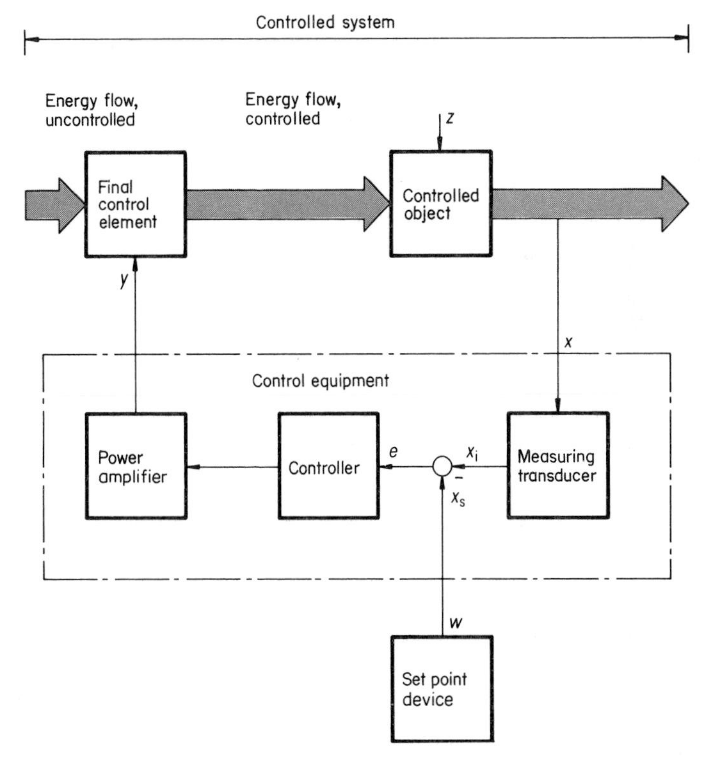

Fig. 8.3
Units of the closed-loop control circuit

The final control element thus acts as an executive element of the control equipment.

8.2.2 Controlled object

The controlled object as a main part of the controlled system is influenced by the final control element. With rectifiers, for example, the d.c. output acts as the controlled object. This is acted upon by disturbances z, e.g. changes in load. The output quantity of the controlled system is picked up by the measuring transducer as a controlled variable and fed to the set/actual comparison in the controller.

8.2.3 Control equipment

The control equipment consists of the measuring transducer, controller and usually the power amplifier. It generates the correcting variable from the set/actual comparison and with it influences the final control element. The set/actual difference, the control difference, is represented by 'command variable minus controlled variable': if the controlled variable x is too small and the difference thus positive, the control equipment must raise the controlled variable by changing the correcting variable y.

If, on the other hand, the controlled variable x is too large, the difference is negative and the control equipment must reduce the controlled variable x by changing the correcting variable y.

The condition is called corrected when the control difference has reached the value zero as closely as possible. How accurately the controlled variable assumes the value of the command variable, i.e. how closely the control difference approaches the value zero, depends on the controller selected.

The *measuring transducer* measures the controlled variable and converts it so that it can be processed as an actual value in the controller's input. With rectifiers the measuring transducers are usually sensors and measuring amplifiers. Measuring shunts, d.c. instrument transducers, current transformers or Hall generators are called sensors. Measuring amplifiers are required in order to arrive at an actual value that can be processed in the controller.

The *controller* consists of a comparator and variable-gain amplifier with wiring for proportional and time response. With rectifiers voltage and current controllers are usually used with an integral component. In the comparator the possibly varying actual value x_i for the controlled variable x is compared with the usually constant set point value x_s for the command variable w.

The command variable is set to a fixed value in the set point device, normally with the aid of a potentiometer, and fed from an outside source to the control equipment.

The stability of this command variable is especially important. For this reason the potentiometer is supplied from a constant voltage source with the least possible temperature variation.

The variable-gain amplifier, also called the operational amplifier, has to generate the desired control response with the aid of its feedback wiring. The output signal of the variable-gain amplifier is usually in the range of ± 10 V. There are final control elements which do not respond with this voltage at a current strength of 5 or 10 mA, in which case *power amplifiers* are also required.

8.3 Operational Amplifiers in the Controller

The operational amplifier or variable-gain amplifier in the controller must be wired for feedback and at its input must compare in a comparator the command variable (set point value) with the controlled variable (actual value), thereby forming the control difference (Fig. 8.4).

A set voltage U_s usually firmly set by the command variable w reaches the comparison point, at which the control difference e is worked out, via the set point channel. The set point value voltage is compared with the actual value voltage $-U_i$ of the controlled variable x. The actual value voltage $-U_i$ is picked up at the output of the controlled object and thus at the controlled system.

The controller is to hold the controlled variable at the normally fixed set value for the command variable. The feedback wiring (here negative feedback) is present so that there is optimum transient response in the control operations.

If a control difference varying from zero occurs, the controller delivers a modified correcting variable y to its output. This is the instruction for the final control

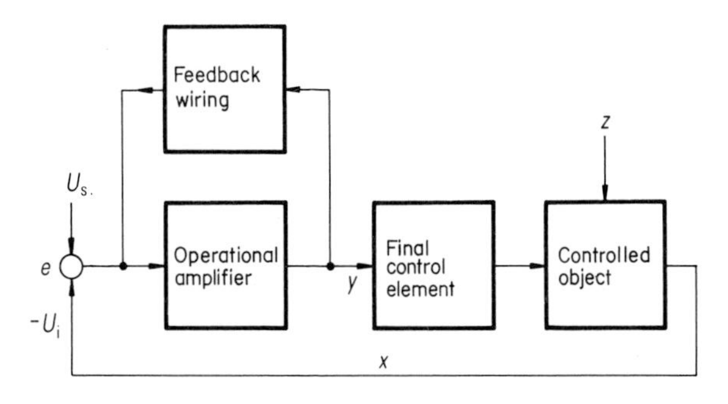

Fig. 8.4
Working of the closed-loop control circuit with the operational amplifier in the controller

element to set its output quantity so that the controlled variable is corrected as quickly as possible to the value specified by the command variable. To obtain the simplest possible control circuit the inverting input of the operational amplifier is used for cascading the control difference. This is the inverting amplifier circuit (cf. Fig. 7.78, Section 7.7.1).

Figure 8.5 shows an inverting amplifier in the controller. The feedback resistance R_f can be seen in the feedback wiring.

The input wiring of the controller consists of the controlled variable channel with resistance R_i and the command variable channel with the resistance R_s. The two channels combine at the inverting input terminal E_- of the operational amplifier. The non-inverting input terminal E_+ of the operational amplifier is connected with the reference potential $M = 0$ V. This connection is usually made via a resistance R_M. The command variable – its reflexion is the set point of voltage U_s – can be set with the potentiometer R. It is fed to the comparison point in the form of a current I_s via the resistance R_s.

The controlled variable – its reflexion is the actual voltage value $-U_i$ – can also be adjusted in the case of many rectifiers. It also reaches the comparison point, where it is compared with the command variable, in the form of a current via resistance R_i. The reflexions of the command variable and controlled variable in the stationary state must always be of different polarity so that the currents formed from the voltages at resistances R_s and R_i generate a differential current I_0, which reflects the control difference. Comparing the currents at the comparison point has the advantage that the voltage reflecting a quantity can always be applied unbalanced to the reference voltage $M = 0$ V. This enables any number of input quantities to be compared.

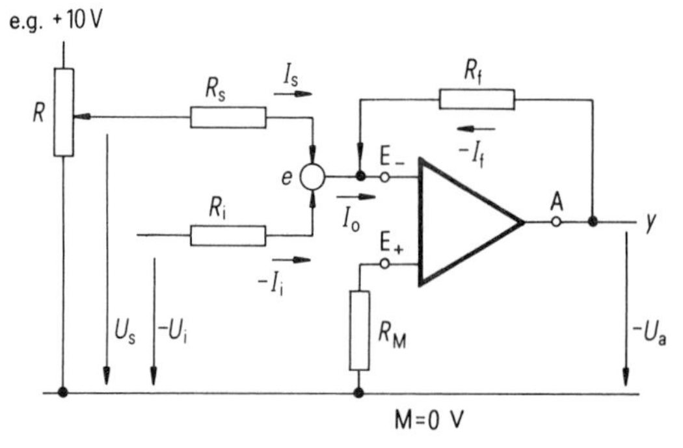

Fig. 8.5 Controller with input and feedback wiring

The constantly adjusted set voltage U_s for the command variable is positive in the stationary state and the actual value voltage $-U_i$ for the controlled variable changing due to disturbances is negative. The set current I_s flows to the comparison point via resistance R_s. Also flowing there is the actual value current $-I_i$ via resistance R_i. At the comparison point there is produced from the two currents the current $I_0 = I_s - I_i$ reflecting the control difference. As the current I_0 (at the inverting input E_-) flows to the highly resistive input of the operational amplifier, the circuit is closed via the feedback resistance R_f.

Only a very small current I_0 can flow off to potential M via the highly resistive internal resistance of the operational amplifier and the non-inverting input E_+. Thus the output voltage U_a of the controller becomes negative when the control difference is reflected by a current I_0 flowing towards the inverting input E_-:

$$\frac{U_s}{R_s} - \frac{U_i}{R_i} = I_0;$$

$$I_0 R_f = -U_a.$$

The output voltage U_a of the controller is the correcting variable y with which the controller intervenes in the controlled system to bring the controlled variable to the value specified by the command variable. The magnitude of the controller's output voltage must possibly be limited to prevent the controlled system being overridden. In many operational amplifiers two inputs are used for this (not shown in Fig. 8.5), to which a positive and a negative limiting voltage are applied. If necessary, these voltages are set with a potentiometer for each of them.

If the set point value equals the actual value ($x_s = x_i$), the correction has been made. There is neither a deviation x_w nor a control difference e ($x_w = 0$ and $e = 0$, shown in the centre of Fig. 8.6). In this case the correcting variable remains constant until

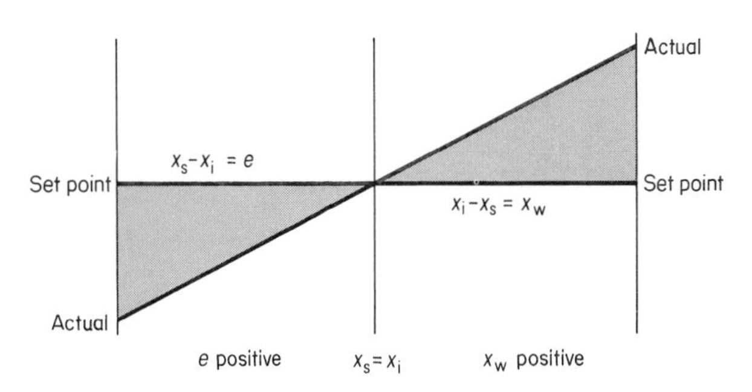

Fig. 8.6 Control difference and deviation

a renewed control difference, i.e. a new deviation in the controlled variable, makes correction necessary again.

As we know, the difference 'command variable minus controlled variable' is called the control difference. 'Controlled variable minus command variable', on the other hand, is called the deviation. The control difference e is positive when the set point value is greater than the actual value ($x_s > x_i$). There is a positive deviation x_w (right part of Figure) when the actual value is greater than the set point value ($x_i > x_s$).

8.3.1 Classification of controllers

A distinction is made between *continuous* and *discontinuous controllers*, e.g. on–off controllers (Fig. 8.7).

There are three basic types of continuous controllers:

P controller (proportional response),
I controller (integral response), also called reset or floating controller,
D controller (differential response), also called derivative controller.

Chiefly P controllers, I controllers and PI controllers (proportional plus integral controllers) are used in a telecommunications power supply system. PD controllers (proportional plus differential) and PID controllers (proportional plus integral plus differential) are also possible.

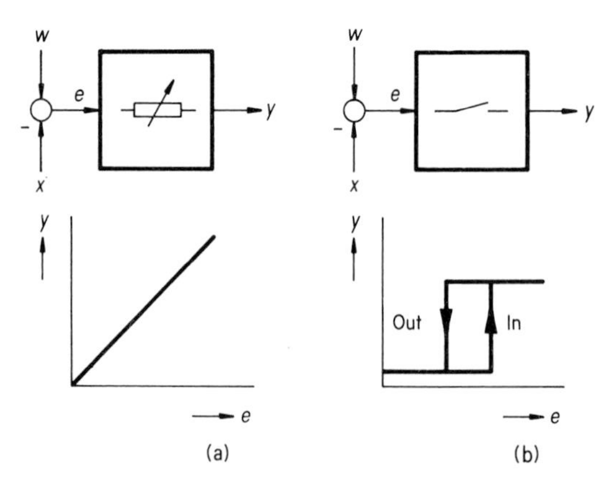

Fig. 8.7 a) Continuous controllers and b) discontinuous controllers – here on–off controller

With the P controller the output quantity (correcting variable y) responds proportionally to the control difference e on the input side. At equilibrium a certain value for the correcting variable is assigned to each value for the control difference (Fig. 8.8). For a correcting variable other than zero to occur, the control difference must be unequal to zero, i.e. the controlled variable does not then have the value specified by the command variable.

The P controller has an ohmic resistance R_f as its feedback (Fig. 8.8a). There is also a potentiometer R in the feedback wiring, with which the proportional factor K_p (also called the proportional amplification V_R) can be set.

At this point it must be noted that the time response at the output of a transfer element to a step-shaped input quantity is called a *step response.* When this is related to the height of the input step it is called *transient response.*

The transient response of the ideal P controller can be seen in Fig. 8.8(b). Here the output voltage of the P controller follows the voltage jump ΔU_0 at the output without inertia at time t_0. The voltage ΔU_0 is to be seen as a reflexion of the control difference.

The following applies to proportional amplification:

$$V_R = \frac{U_a}{\Delta U_0} = \frac{R_f}{R_0}.$$

Thus proportional amplification of, for example, $V_R = 3$ is obtained when the feedback resistance R_f is three times greater than the input resistance R_0. The output

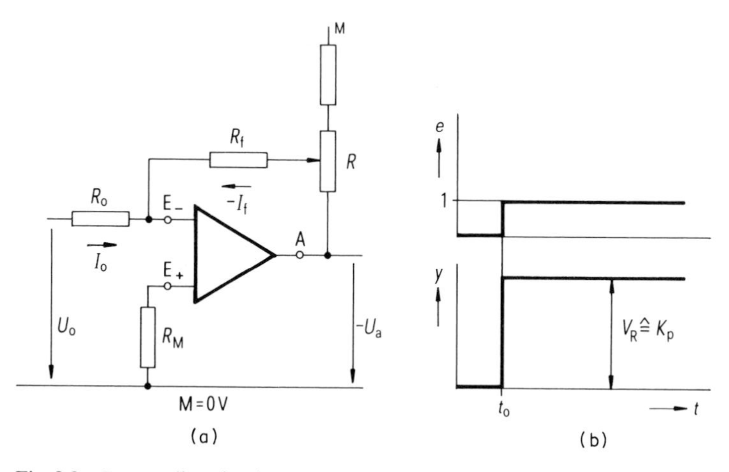

Fig. 8.8 P controller, circuit and transient function

voltage U_a is then three times larger than the input voltage U_0. The example in Fig. 8.8(b) is based on amplification of $V_R = 3$.

P controllers are characterized by *rapid correction* of the disturbance (good initial response). However, a residual error has to be accepted, for the output quantity of the P controller is always proportional to the remaining control difference (poor correction response).

I controller

With the I controller a certain rate of change of speed for the correcting variable y is assigned to each value for the control difference e (Fig. 8.9).

The I controller shifts its output quantity in the direction for correction until the control difference is practically zero. It has a capacitor C_f as feedback (Fig. 8.9a). In the ideal case the transient function of an I controller is the same as the diagram in Fig. 8.9(b). The feedback capacitor C_f determines, together with input resistor R_0, the time response of the I controller. If the input quantity, in Fig. 8.9(a) the voltage U_0, is equal to zero, the feedback capacitor C_f remains in the charging state it has reached.

The output voltage U_a remains at the value it has reached, normally a value other than zero. If the control difference e assumes a value different from zero, the output voltage U_a changes in the opposite direction to the control difference at a speed determined by the integrating factor:

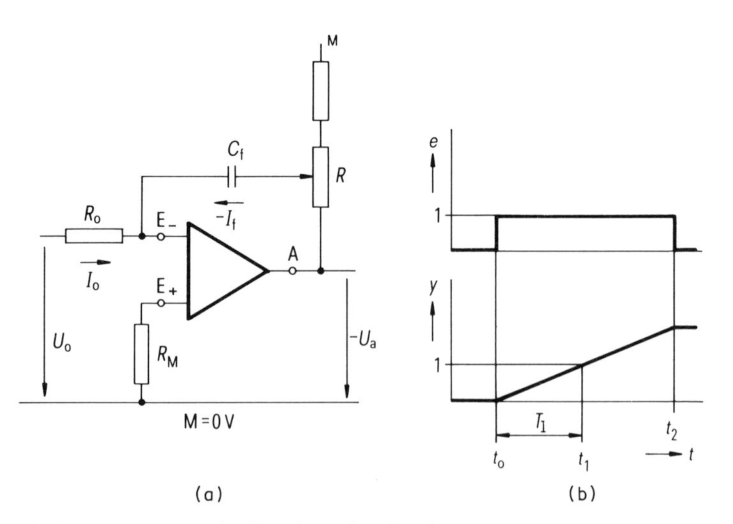

Fig. 8.9 I controller, circuit and transient function

$$K_I = \frac{1}{R_0 C_f}.$$

This means that the output voltage U_a changes relation to the initial value U_{a0} as shown by the equation:

$$-U_a(t) = \frac{1}{R_0 C_f} \int_0^t U_0(t) dt + U_{a0}.$$

If at time t_0 the input voltage U_0 makes a jump, the integrating time ($T_I = R_0 C_f$) is that time during which the output voltage U_a, starting from the initial voltage U_{a0}, undergoes a change; it depends on the height of the voltage jump on the input side (time t_1).

The output voltage U_a changes continuously as long as there is a voltage at the input differing from zero. If the input voltage again becomes zero (time t_2), the output voltage remains at the final value it has reached.

The I controller permits very *accurate correction* of the disturbance (good correction response), as it is only when the control difference has become zero that the output quantity no longer changes any more. One disadvantage is that it intervenes relatively slowly (poor initial response) because the control difference must first be integrated and a sufficient correcting variable built up.

The integrating time can be set with the potentiometer R when the input resistance R_0 is very large compared with the resistance R of the potentiometer.

PI controller

With the PI controller the correcting variable y is the same as the sum of the output quantities of a P and an I controller (Fig. 8.10).

The PI controller initially responds like an I controller, i.e. any control difference results in integration of the same so that the output quantity moves towards the set point. In addition and simultaneously with the control difference, a proportional response is superimposed on the integral one until the control difference, except for the control error particular to the operational amplifier, has been eliminated. This means that the PI controller does more than the I controller. With the aid of the P component it forms a derivative action, the P derivative action.

The PI controller has in its feedback wiring a resistor R_f and a capacitor C_f (Fig. 8.10a). This controller combines the advantages of both controllers, namely the *rapid reaction* of the P controller with the *accuracy of correction* of the I controller, without displaying the disadvantages of either.

The transient function of the PI controller in the ideal case is as shown in Fig. 8.10(b). If at time t_0 the input voltage U_0 intended as a reflexion of the control

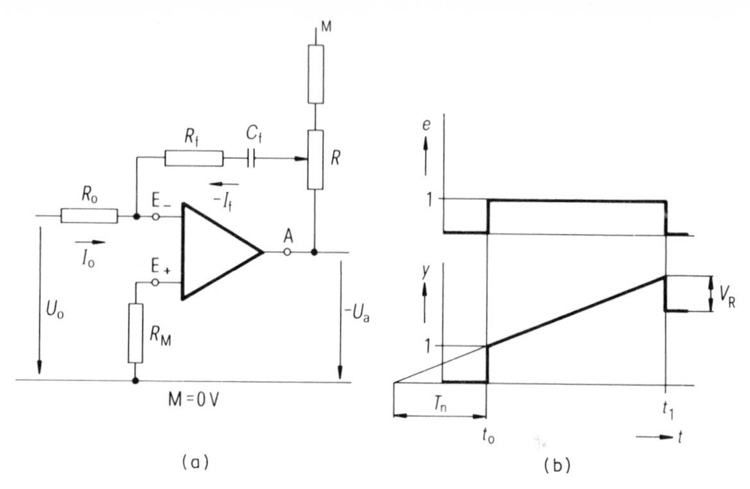

Fig. 8.10 PI controller, circuit and transient function

difference makes a jump from zero, the output voltage U_a also makes a jump from the level $\Delta U_0 V_R$. As the capacitor C_f acts as a short-circuit, only the input voltage U_0 and the proportional amplification V_R are decisive for the output voltage U_a. The output voltage U_a then alters as a straight line in time from the voltage jump made in accordance with the integration time $T_I = R_0 C_f$. The rise in output voltage results from the magnitude of the charging current $I_f = \Delta U_0 / R_0$ and according to the capacitance of the capacitor C_f.

If the input voltage U_0 at time t_1 returns to zero, the P derivative action disappears and the PI controller jumps down on the output side, similarly from the height $\Delta U_0 V_R$. The output voltage of the PI controller then remains at the value there would have been had only the I component been effective.

As with the P controller, the proportional factor (the magnitude of proportional amplification V_R) is set with the feedback potentiometer R. The proportional amplification V_R of a PI controller indicates, for each jump in the input quantity, the factor for the change in input quantity as referred to the output voltage jump in the first instance.

Integration or reset time T_n is the name given to that time by which the I controller would have to intervene earlier to achieve the same change in correcting variable as a PI controller. With the start of the signal jump at the input the PI controller has requested at its output an input signal V_R times too much, but at the end of the signal jump has reduced the request by the same amount. With the reset time T_n the output signal is adjusted, as if the PI controller had started integrating earlier by the time T_n.

8.4 Controlled Rectifiers with Thyristor Power Section

Figure 8.11 shows the construction of a controlled rectifier with a thyristor power section and phase-angle control. Such rectifiers are suitable mainly for the middle and upper power range.

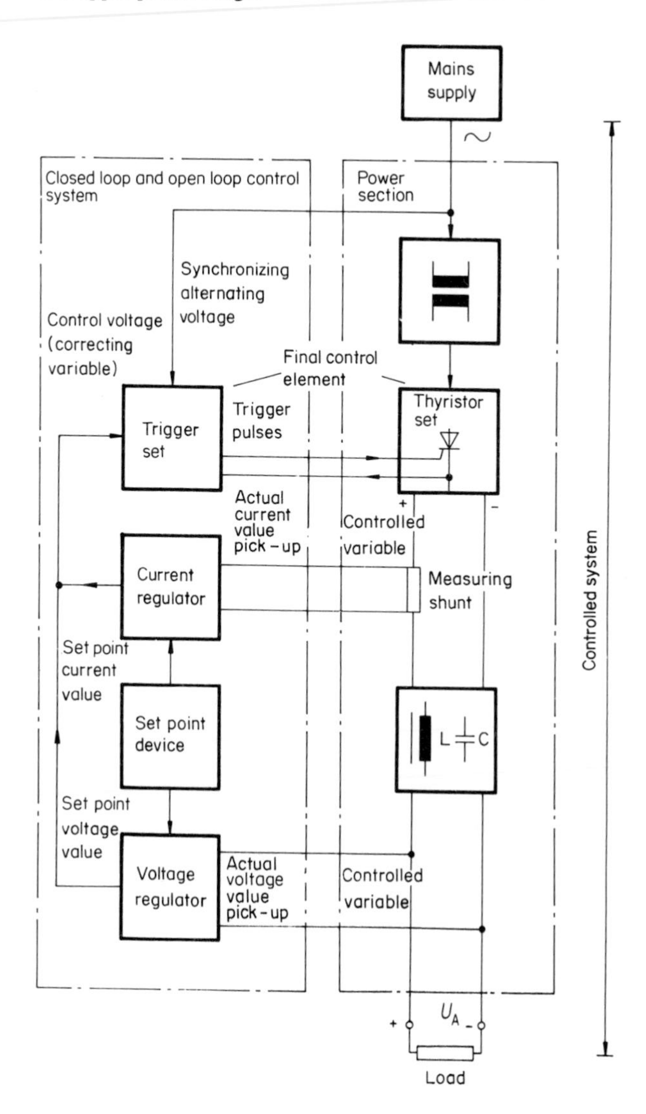

Fig. 8.11
Block diagram of a controlled rectifier with thyristor power section and phase-angle control

8.4.1 Power section

The alternating supply voltage is transformed by the *main transformer* so that the maximum required d.c. output voltage of the rectifier can be obtained. The main transformer also isolates the mains supply and load electrically.

An auxiliary transformer (not shown in Fig. 8.11) provides the synchronizing alternating voltage from the supply voltage.

The thyristor set together with the trigger set acts as a *final control element*. It has the task of providing the required direct voltage. From the measuring shunt the actual current value reaches the current regulator as voltage.

The direct voltage is smoothed by the *filters*, i.e. filtered to the permissible level of superimposed alternating voltage (cf. Section 2.3.1).

The direct voltage U_A is tapped at the output terminals of the power section.

8.4.2 Closed-loop and open-loop control system

The "closed-loop and open-loop control" assembly combines all the essential functions used in forming the trigger pulse. There are different versions of this assembly depending on whether it is connected to an alternating current or a three-phase supply. Figure 8.12 shows the design for a three-phase unit.

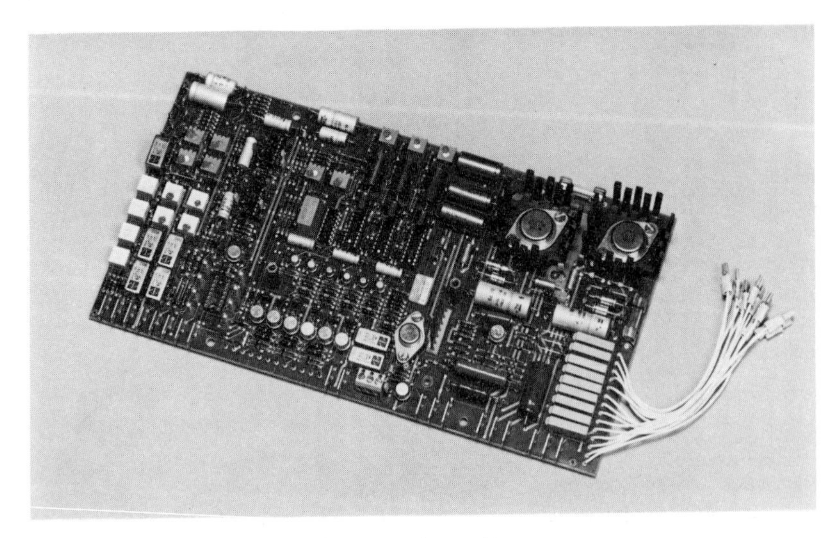

Fig. 8.12 Closed-loop and open-loop control assembly

Set point device

The set point for the direct voltage to be controlled is set in the set point device with the aid of a potentiometer. It also has a potentiometer for setting the current limit.

Voltage regulator

The direct voltage to be controlled is tapped at the output of the power section and fed to the voltage regulator (controller) as an actual voltage value (controlled variable). The set point voltage value passes from the set point device so that there is a 'set voltage value/actual voltage value' comparison at the voltage regulator's input. The resultant control difference is processed by the voltage regulator so that through its proportional and time response a correcting variable is produced, which with the aid of the trigger set determines a suitable firing time for the thyristor set.

Trigger set

The trigger set supplies the trigger pulses for the thyristors in the thyristor set. It adjusts the output voltage of the thyristor set according to the correcting variable, i.e. the output voltage from the voltage regulator.

To do this, depending on the supply voltage, a sawtooth voltage is generated from the synchronizing alternating voltage which starts with each zero passage of the phase. The output voltage of the voltage regulator, the control voltage U_{St}, is compared with the sawtooth voltage. If equal, a trigger pulse is passed to the thyristor set. The direct voltage at the output of the thyristor set is determined by shifting the firing time.

Current regulator

The current regulator is used to protect the rectifier against overloading. The maximum permissible current is set with the aid of a potentiometer in the set point device. The current regulator only intervenes in the control circuit if the actual current value wants to become greater than the set, maximum permissible, set point current value. In a case of limiting the current the current regulator reduces the d.c. output voltage until the rectifier cannot deliver more than the set current (this is usually the rated current 100% I_{rated}, cf. Fig. 8.14).

The *control processes* in four instances are explained below (Figs. 8.11 and 8.13):

① d.c. output voltage constant,
② d.c. output voltage too high,
③ d.c. output voltage too low,
④ current limitation.

① Let the set point value for the d.c. output voltage U_A of the rectifier be, for example, 51 V. If the d.c. output voltage is now actually 51 V, the voltage regulator detects no difference between the set voltage value and the actual voltage value. The control difference is zero. Thus the control voltage U_{St} at the output of the regulator remains constant and the trigger set gives, in relation to the zero passage of the supply voltage, the trigger pulse to the associated thyristor at the same time as with the preceding voltage half-waves. The control angle may, as in Fig. 8.13(a), lie, for example at about 80° in relation to the zero passage. The d.c. output voltage in the given example therefore remains constant at 51 V.

② With a decreasing, purely resistive load the output voltage of the rectifier might initially become a little larger. In the example U_A becomes >51 V. As soon as the actual voltage value becomes greater than the set voltage value there is a negative control difference (positive deviation). This results in the voltage regulator delivering a somewhat higher control voltage U_{St}. Following this the trigger pulses from the trigger set are emitted a little later to the respective thyristor, i.e. shifted in the direction of 180° (Fig. 8.13b, cf. Section 7.6.3). This means later firing of the associated thyristor so that the d.c. output voltage of the rectifier drops and returns to the set point of 51 V.

③ With an increasing, purely resistive load the output voltage of the rectifier initially falls a little below the value of 51 V because a somewhat higher source voltage is now required throughout the whole d.c. circuit to achieve the required set voltage of 51 V for the load. Consequently, the source voltage must be adjusted somewhat higher by the control process. As the actual voltage value in the initial

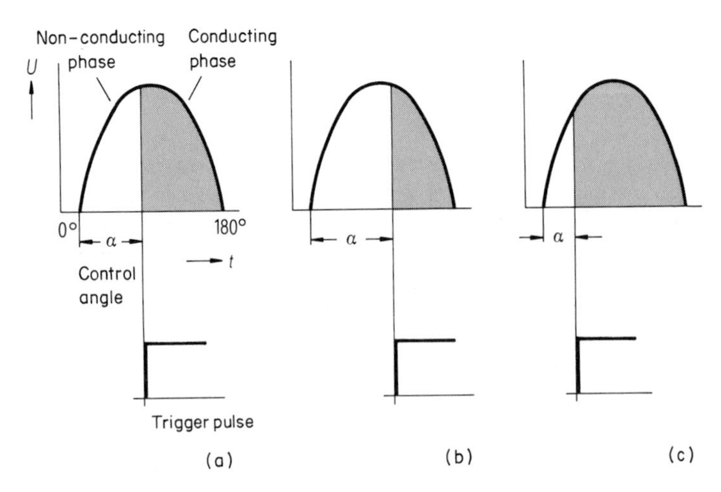

Fig. 8.13 Shift of trigger pulses

moment of stepping up the load drops slightly below the set voltage value, there is a positive control difference. It follows from this that the voltage regulator delivers at its output a slightly lower control voltage U_{St} to the trigger set. The trigger set therefore now generates trigger pulses for the thyristors which fire the thyristor a little earlier in relation to the voltage zero passage of the half-wave. The trigger pulses are thus shifted a little in the direction of 0° (Fig. 8.13c). In this way the d.c. output voltage of the rectifier again rises to the set point (51 V).

④ The control processes considered thus far (examples ①, ② and ③) are executed by the voltage regulator alone without the current regulator having to intervene, because the loading (currentwise) lay within the permissible range.

Example ④ now concerns the current regulator. Let the current limit be set at 100% I_{rated}. As long as the actual current value is less than the set current value coming from the set point device, the current limiter does not come into play. If, however, the actual current value starts to become greater than the set current value, the current regulator cuts out the voltage regulator by delivering the now higher control voltage U_{St}.

Trigger pulses are now generated by the trigger set that are so far shifted towards 180° that a d.c. output voltage is produced at the rectifier which drives only the maximum permissible current, thereby protecting the rectifier against overloading (cf. Fig. 8.14).

8.4.3 Behaviour of the output voltage and dynamic response

Figure 8.14 shows the behaviour of the *d.c. output voltage* of a thyristor-controlled rectifier, indicating the static tolerance range and the current limitation. The static tolerance range means the range for the controlled variable within which it must remain after a control process. The static tolerance for rectifiers with a thyristor power section is, for example, ±0.5% (cf. Section 2.2).

If changes in disturbance (here primarily loading of the rectifier) occur in closed-loop control, the control equipment reacts with a transient of the controlled variable characterizing the *dynamic response* of the control circuit (Fig. 8.15). With stepped changes in load this can result in short-time deviations of the d.c. output voltage. Depending on the extent of the change in load the output voltage will lie more or less far outside the static tolerance range with one or more half-waves of the transient. There are two comparative values for judging the control response when correcting a stepped disturbance. These are transient overshoot and settling (correction) time.

The *transient overshoot* denotes the amplitude of the first deflection of the controlled variable when settling the disturbance.

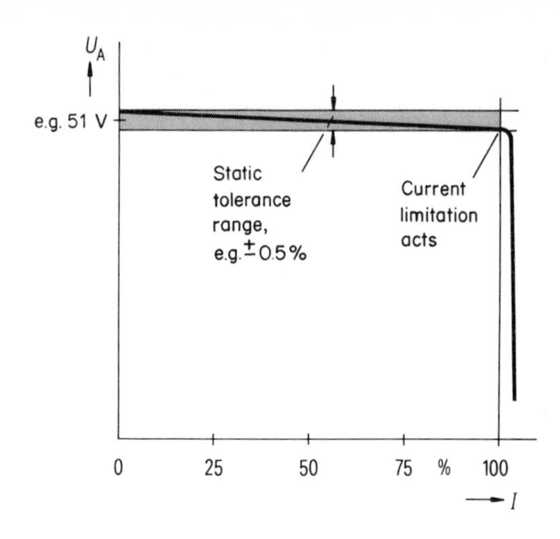

Fig. 8.14 Behaviour of the d.c. output voltage

By *settling time* is understood the time that passes after the occurrence of a stepped change in load until the d.c. output voltage returns to the static tolerance range. At time t_1 there occurs a stepped change in the load in the negative direction. The d.c. output voltage will then initially experience a positive deflection. There now is an appreciable control difference and the regulator returns the voltage to the set point (e.g. 51 V).

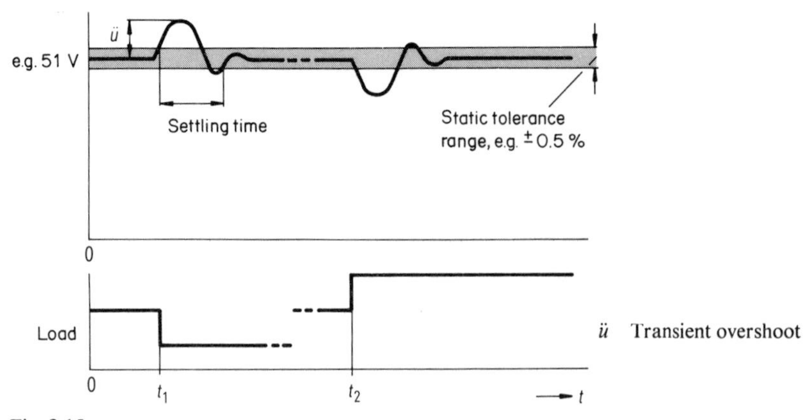

ü Transient overshoot

Fig. 8.15
Behaviour of d.c. output voltage after a stepped change in load

At time t_2 there occurs a stepped change in load in the positive direction. The control process now takes place in the opposite direction. The deflection of the voltage to be controlled is here in the negative direction. A deviation of $\pm 4\%$ from the desired d.c. output voltage is the permitted tolerance range.

8.5 Controlled Rectifiers with Thyristor Power Section and Switching Controller

Power transistors or thyristors act as the final control element when using switching controllers for controlled rectifiers. As thyristors are used only rarely, there will only be a short explanation here of the working of the switching controller with thyristors.

Figure 8.16 shows the schematic circuit diagram of a voltage closed-loop control system with a switching controller (using a thyristor). Rectifiers constructed on this principle are suitable for the lower to medium power range.

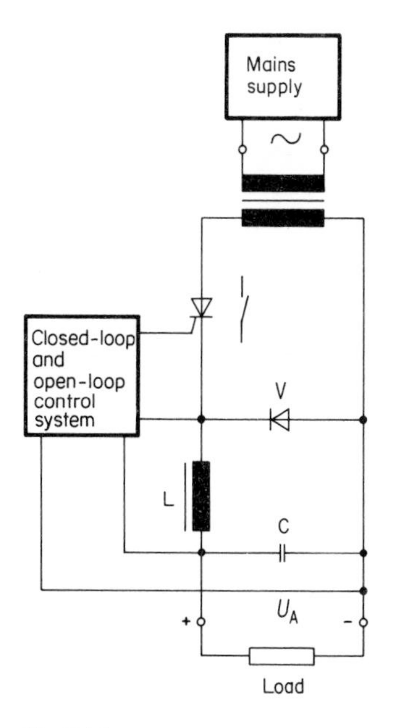

Fig. 8.16
Controlled rectifier with thyristor power section and switching controller

The closed-loop and open-loop control assembly delivers as a function of the d.c. output voltage trigger pulses in the positive half-wave of the alternating supply voltage for the thyristor in order to charge the charging capacitor C up to the voltage set in the set point device via the thyristor in the on condition. The smoothing choke L is used to avoid steep current rises at the capacitor C. The respective trigger pulse appears earlier or later during the positive voltage half-wave, depending on the discharge from capacitor C due to the load. As the smoothing choke has stored energy during the charging current surge, a 'free-wheeling diode' V, via which the choke's energy can flow off, is necessary for the time in which the thyristor is in the off condition.

8.6 Controlled Power Supply Equipment with Transistor Power Section

8.6.1 Controlled rectifiers with longitudinal controller

Figure 8.17 shows a controlled rectifier with a transistor longitudinal controller. The power transistor is variably controlled by the closed-loop and open-loop control assembly, depending on the level of the measured d.c. output voltage compared with the set voltage: if the direct voltage at the output of the rectifier V is appreciably higher than the d.c. output voltage U_A to be provided, a voltage drop can be set at this transistor of such a size that a constant output voltage is produced.

The efficiency that can be achieved with this circuit is relatively low. Favourable factors are rapid correction and the particularly simple circuit design. Controlled rectifiers with a transistor longitudinal controller are used to advantage in the lower power range, up to about 50 W.

The longitudinal controller principle is often found in power supply equipment for providing internal supply voltages, for example, integrated circuits in power supply subassemblies.

8.6.2 Controlled rectifiers with switching controller

Figure 8.18 shows a controlled rectifier with switching controller. A supply-side bridge circuit V2 supplies the 'intermediate circuit voltage'. According to the actual value voltage ($\widehat{=} U_A$) measured, the closed-loop and open-loop control assembly in each cycle switches the switching transistor to the conducting state long enough for the mean voltage value of that cycle to match the set point voltage value for the d.c. output voltage U_A.

The clock frequency is frequently 20 kHz or more. The human ear is insensitive to this frequency.

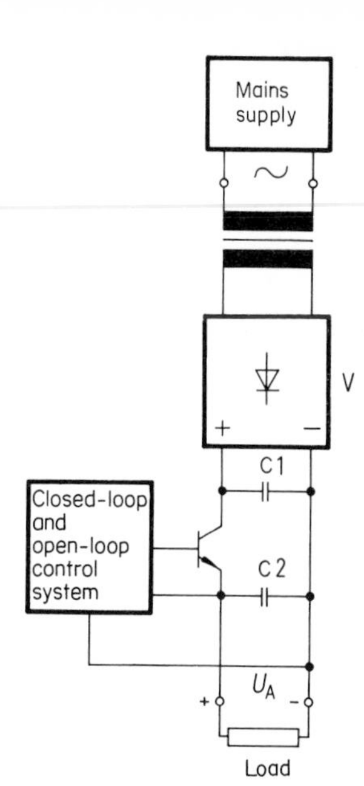

Fig. 8.17
Controlled rectifier with transistor power section and longitudinal controller

With the switching transistor in the conducting state current flows from the positive pole V2 to the switching transistor, via the smoothing choke L to the charging capacitor C2 and via the load to the negative pole V2. When the conducting phase for the switching transistor has ended the current continues flowing via the smoothing choke L and the free-wheeling diode V1, capacitor C2 being discharged.

As the clock frequency is relatively high, the filtering section must be kept small. However, the high frequency does require certain measures for the suppression of radio interference. Figure 8.18b shows the voltage at the diode V1 dependent on the time.

In example 1 shown in Fig. 8.18 it is assumed that the 'conducting' state t_1 of the switching transistor is of the same length as the 'non-conducting' state t_2. With a change in load or in the intermediate circuit voltage either the period t_1 becomes longer or the period t_2 shorter (example 2) or the period t_1 shorter and the period t_2 longer (example 3). The period T remains constant. This is called the *pulse-width control* (cf. Section 7.6.4). The longer the conducting state t_1 lasts and thus the larger the voltage–time surface, the higher the average value for the d.c. output voltage.

8.6.3 Switching-mode power supplies

The principle described in connection with the switching controller can also be applied to switching-mode power supplies.

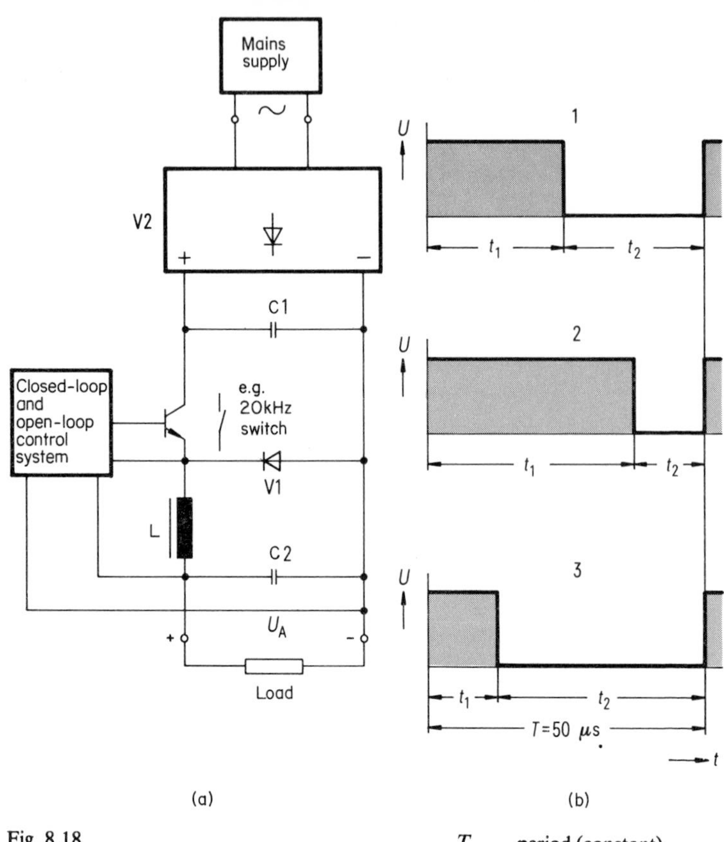

Fig. 8.18
Controlled rectifier with transistor power section and switching controller

T period (constant)
t_1 conducting phase
t_2 non-conducting phase

As shown in Fig. 8.19, a switching-mode power supply consists of a power section and the closed-loop and open-loop control assembly. The power section can be broken down into a supply-side rectifier and a d.c./d.c. converter. The supply-side rectifier supplies the voltage for the d.c./d.c. converter (intermediate circuit voltage).

The alternating voltage from the supply mains is rectified (V1), filtered (C1) and then 'chopped' by a switching transistor. The resultant alternating voltage (rectangular shape) is again rectified after transforming. The transformer is also used for electrical separation. The rated constant d.c. output voltage U_A is then available after filtering.

The closed-loop and open-loop control assembly has to keep the d.c. output constant. Pulse-width control is also used here, as with the switching controller. Switching-mode power supplies are suitable for the lower (>50 W) to medium power range. Figure 8.20 shows a view of a switching-mode power supply.

There are 'primary cycled' and 'secondary cycled' switching-mode power supplies. Primary cycled ones, which are the type used almost exclusively today, are based on the blocking converter and flow converter principle.

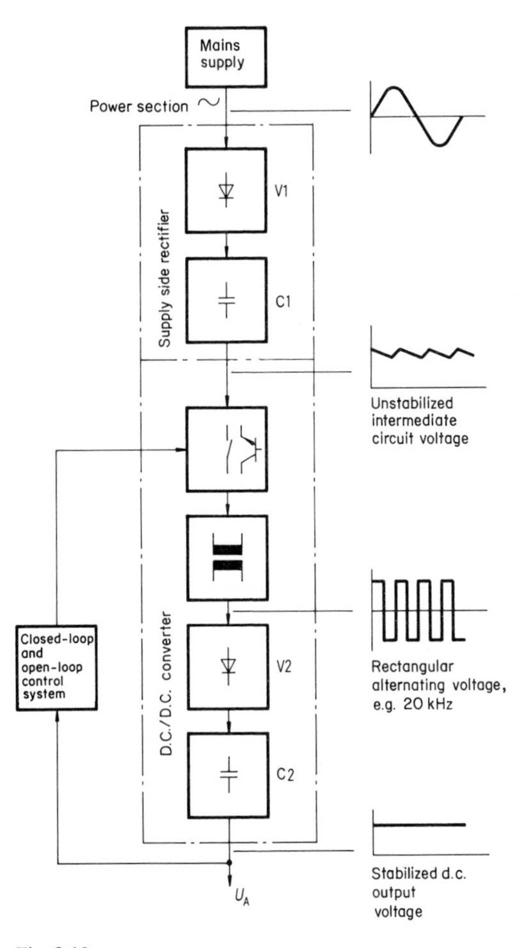

Fig. 8.19
Principle of a switching-mode power supply

Fig. 8.20 Switching-mode power supply 48 V/30 A

8.6.4 D.C./D.C. converters

Single-ended flow converter

With the single-ended flow converter the consumption of energy on the primary side coincides in time with the delivery of energy on the secondary side. Figure 8.21 shows a single-ended flow converter with a switching transistor. Here the capacitor C1 and the inductance of the transformer form the 'primary-side resonant circuit'. A second resonant circuit ('secondary-side resonant circuit') is represented by the winding c and capacitor C2. Current flows in the primary-side resonant circuit during the conducting phase of switching transistor V3 which leads to current flowing to the load in the secondary-side resonant circuit via the now-conducting diode V1. Capacitor C2 is at the same time being charged. This is possible because

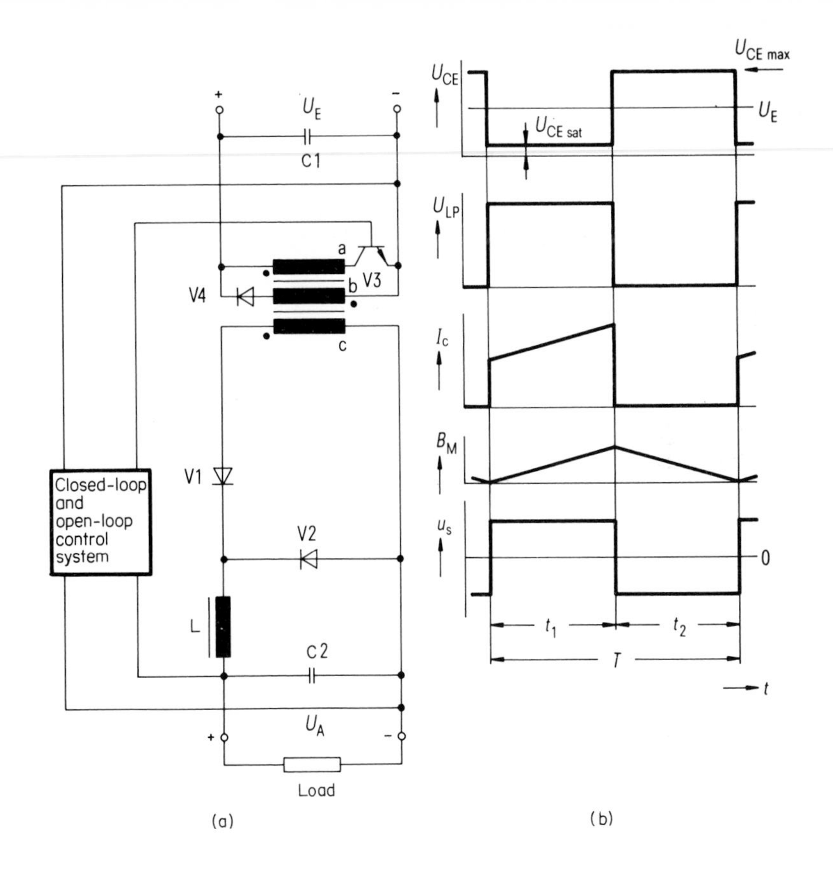

U_{CE}	Voltage between collector and emitter of transistor V3
$U_{CE\,sat}$	Saturation voltage
$U_{CE\,max}$	Maximum voltage between collector and emitter
U_{LP}	Voltage at the primary winding (a) of the transformer
I_c	Collector current (primary current)
u_s	Secondary voltage
T	Period
t_1	Conducting phase transistor V3
t_2	Non-conducting phase transistor V3

Duty cycle $V_T = 0.5$

Fig. 8.21
Single-ended flow converter with one switching transistor

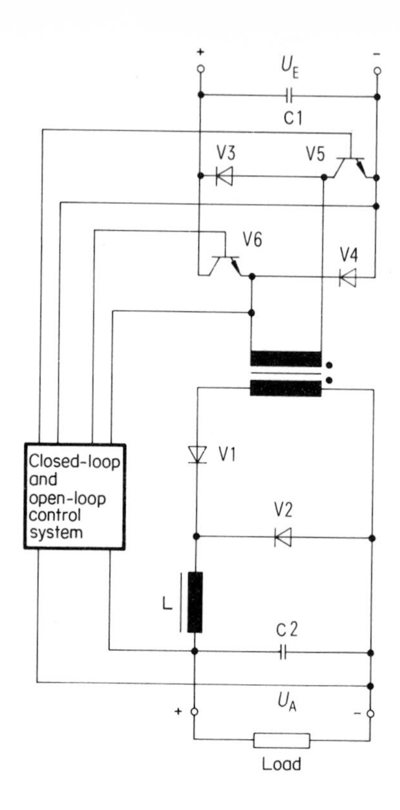

Fig. 8.22
Single-ended flow converter with two switching transistors

the primary and secondary windings have the same sense of winding (marked by a dot on the same side at both windings a and c). The smoothing choke L stores energy during this process.

During the non-conducting phase of switching transistor V3 the supply of current is interrupted in the secondary circuit. The choke still delivers energy to the load for a short time. Smoothing capacitor C2 is discharged. Free-wheeling diode V2, now in the conducting state, lies in parallel so that the current still continues flowing.

The closed-loop and open-loop control system switches transistor V3 to the conducting state in a period long enough for the d.c. output voltage to be kept constant with a fluctuating input voltage or changing load.

There is a demagnetizing winding b so that the transformer does not become saturated. The primary side oscillation circuit turns over during the non-conducting phase of switching transistor V3. Current flows in the reverse direction via diode V4 and the demagnetizing winding, thereby preparing the transformer for the next conducting phase.

The circuit of the single-ended flow converter with *two* switching transistors shown in Fig. 8.22 is a variant that also occurs in telecommunications power supply systems. Its function is generally the same as that explained in connection with Fig. 8.21. In this case the two diodes V3 and V4 demagnetize the transformer. The important factor with this circuit is that both power transistors V5 and V6 are made to conduct *simultaneously* by the closed-loop and open-loop control system.

Single-ended blocking converter

In contrast with the single-ended flow converter (Fig. 8.21), in the case of the single-ended blocking converter (Fig. 8.23) energy consumption and energy delivery are shifted 180° in time. The inductance of the transformer acts together with the capacitance of capacitor C1 as a parallel resonant circuit. During the conducting phase of switching transistor V3 current flows via the transformer (Fig. 8.23a). Primary and secondary windings have an opposite sense of winding. Thus in the conducting phase of switching transistor V3 the diode V1 is polarized in the reverse direction and no current can therefore flow via the secondary winding. The load is supplied exclusively from smoothing capacitor C2.

In the non-conducting phase of the switching transistor the polarity of the voltage at the transformer reverses. Diode V1 now becomes conducting. The energy stored in the transformer during the conducting phase is delivered to the load. At the same time capacitor C2 is recharged. Diode V2 attenuates induced voltage peaks.

Figure 8.23(b) illustrates the behaviour of the individual variables.

Push–pull flow converter

The push–pull flow converter (Fig. 8.24) is a combination of two single-ended flow converters which within a period are operated by the closed-loop and open-loop control system with a phase shift of 180°. When transistor V3 is conducting, current flows via the primary winding of the transformer P1. This produces a voltage in both secondary windings S1 and S2. As these windings are connected in the opposite direction to primary windings P1 and P2, a voltage of positive polarity is now at the outside of winding S1 so that diode V1 becomes conducting, a current flow thereby being formed in the circuit: Winding S1, diode V1, choke L, load and then back to winding S1. The smoothing capacitor C2 is charged.

At the secondary winding S2, which is also connected in the opposite direction to the primary winding P1, there is now a voltage of negative polarity at the outside so that diode V2 becomes non-conducting.

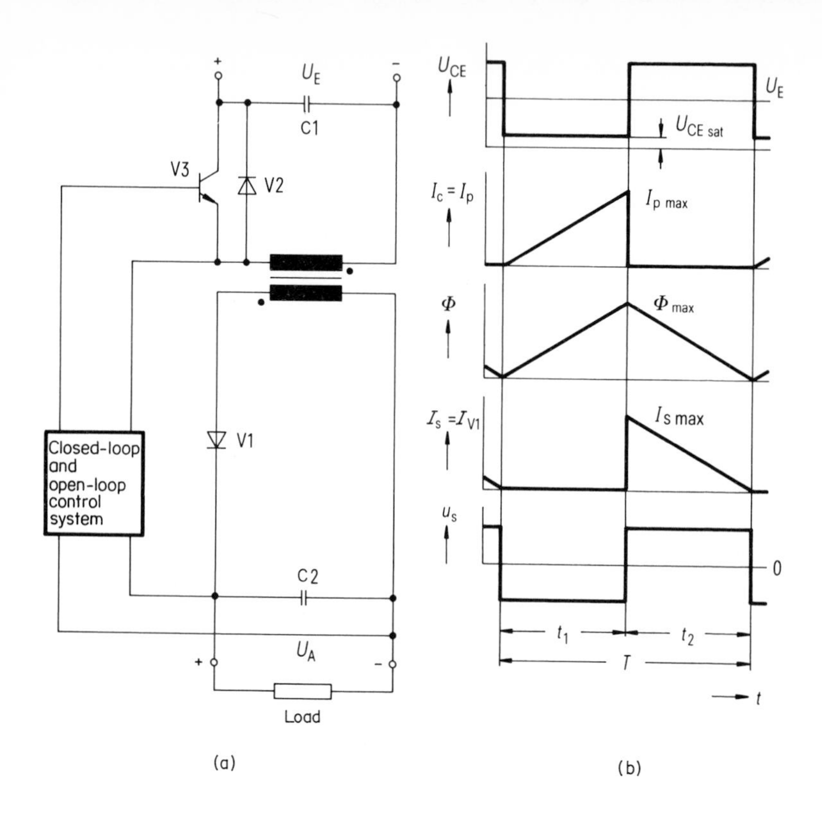

U_{CE}	Voltage between collector and emitter
$U_{CE\,sat}$	Saturation voltage
U_E	D.C. input voltage
I_c	Collector current
I_p	Primary current
$I_{p\,max}$	Maximum primary current
Φ	Magnetic flux
Φ_{max}	Maximum magnetic flux
I_s	Secondary current
$I_{s\,max}$	Maximum secondary current
I_{V1}	Current via diode V1
u_s	Secondary voltage
T	Period
t_1	Conducting phase V3
t_2	Non-conducting phase V3

Fig. 8.23 Single-ended blocking converter

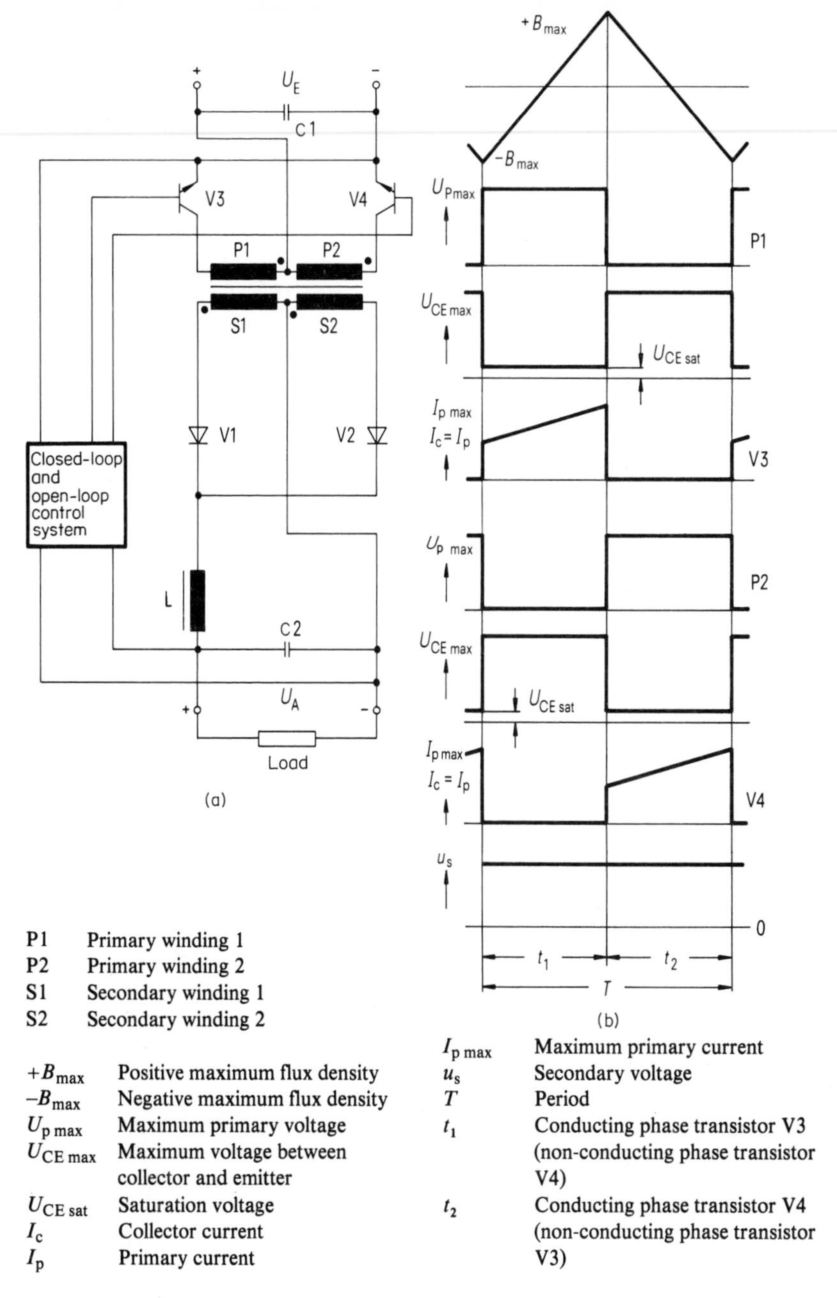

Fig. 8.24 Push–pull flow converter

The duration of the conducting phase of switching transistor V3 is determined by the closed-loop and open-loop control system. Once the conducting phase has ended there follows a state in which both switching transistors V3 and V4 are non-conducting, V3 is already blocked and V4 is not yet conducting. During this time the choke L acts as 'current source'; together with capacitor C2 it continues supplying the load with current (at a constant voltage). Half of the current (load current) flows through each of the two parts of the transformer's secondary winding. The diodes V1 and V2 act here as free-wheeling diodes.

The second half-period runs as a mirror image of the first half-period. First the switching transistor V4 becomes conducting and with it also diode V2. Diode V1, on the other hand, now becomes non-conducting. With the end of the transition time of transistor V4 both switching transistors V3 and V4 are again non-conducting. A new period starts.

The behaviour of the individual variables can be seen from Fig. 8.24(b). There is a conducting and non-conducting phase for each of the two switching transistors within the period T determined by the clock frequency (e.g. at 20 kHz, 50 μs). The conducting phase is always shorter than the non-conducting phase, i.e. also shorter than a half-period.

The start of the conducting phase of switching transistor V3 is shifted 180° in time compared with the start of the conducting phase of switching transistor V4. During the conducting phase of switching transistor V3 (t_1) the magnetic flux density B changes from a negative to a positive maximum value. In the reverse direction the magnetic flux density changes during the conducting phase of switching transistor V4.

It is important for the function that the conducting phases of the two switching transistors do not overlap and for this reason a minimum pulse spacing is provided in practice between the conducting phases of the transistors. This is not shown in the idealized representation in Fig. 8.24(b).

Pulse width control is used for the closed-up control system as with the other types of converters.

Up to now the discussion has been on d.c./d.c. converters for a single output voltage with direct closed-up control. There now follow examples of d.c./d.c. converters for several output voltages with direct and e.m.f. control. These are based on equipment working on the single-ended flow principle.

An example of a single-ended flow converter with *direct control* and *one* output voltage has already been shown in Fig. 8.21. With this equipment it is possible to provide a voltage with narrow tolerance (e.g. with component voltages from 5 to 24 V ±4%) and wide power range.

Figure 8.25 shows a single-ended flow converter which supplies a *master-controlled* output voltage U_{A1} with narrow tolerance and wide power range. The actual voltage value for the closed-loop and open-loop control system is tapped at this output

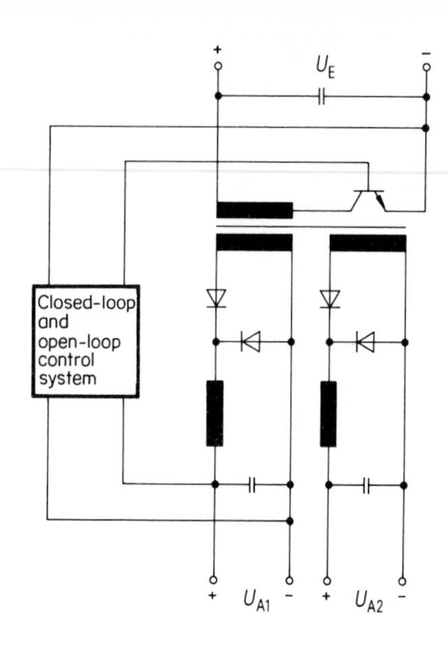

Fig. 8.25
Single-ended flow converter with master-controlled output and one jointly controlled output

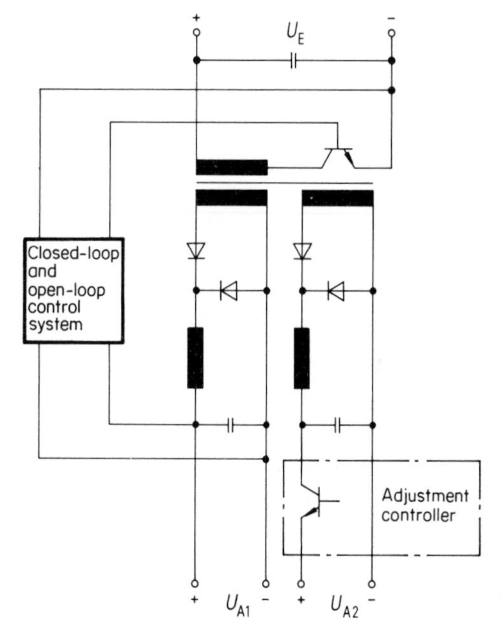

Fig. 8.26
Single-ended flow converter with one master-controlled and one jointly controlled output with adjustment control

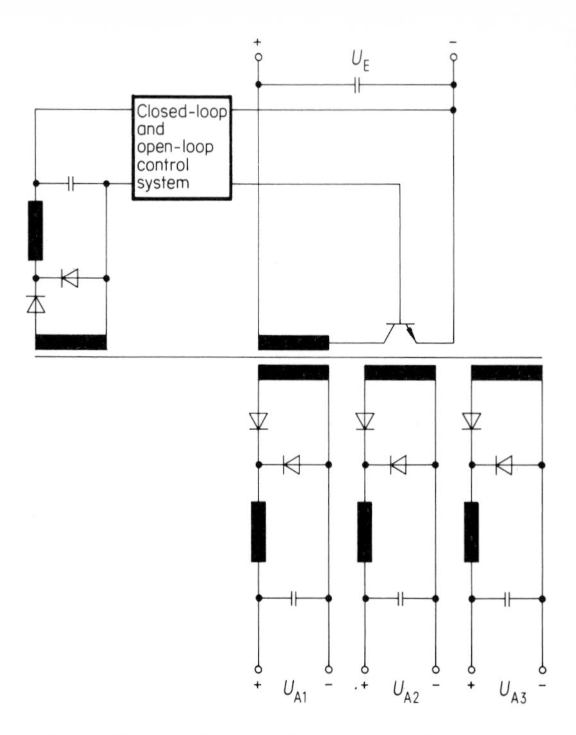

Fig. 8.27
Single-ended flow converter with e.m.f. control

voltage. The *jointly controlled* output voltage U_{A2} benefits from the control to a minor extent.

The level of this voltage depends on the load on the master-controlled output. For this reason it is only possible to achieve a wide tolerance of, e.g., $\pm 10\%$ for the jointly controlled output.

If a narrow tolerance is also required for the jointly controlled output voltage (for smaller powers), a (continuous) *adjustment controller* (*longitudinal controller*) must also be provided (Fig. 8.26).

E.M.F. control (Fig. 8.27) works differently from the principle of direct control of the output voltage shown in Figs. 8.21, 8.25 and 8.26. It is more simple in design that the previously described control systems. The disadvantage here is the somewhat wider tolerance for the output voltage (e.g. $\pm 5\%$ with component voltages from 5 to 24 V).

With this principle the output voltage of an auxiliary winding of the transformer is tapped as the actual value and in the closed-loop and open-loop control system is compared with the single set point of the joint set point device. The wiring of this auxiliary winding must be the same as with the secondary windings for the output voltages U_{A1}, U_{A2}, etc.

Bibliography

Books

Team of authors: *Electronics*, 2nd Part: *Industrial Electronics*, Verlag Europa-Lehrmittel, Wuppertal-Barmen (1970).

Team of authors: *Converter Engineering*, AEG-Telefunken, Elitera Verlag, Berlin (1977).

Benz, W., Heinks, P., and Starke, L.: *Tables for Electronics and Communications Engineering*, Kohl + Noltemeyer Verlag GmbH, Dossenheim, and Frankfurter Verlag, Frankfurt on Main (1980).

Bergtold, F.: *Dealing with Operational Amplifiers*, Verlag R. Oldenbourg, Munich and Vienna (1975).

Beuth, K.: *Components in Electronics; Electronics* 2, Vogel-Verlag, Würzburg (1980).

Ernst, D., and Ströle, D.: *Industrial Electronics; Principles – Method – Applications*, Springer-Verlag, Berlin, Heidelberg, New York (1973).

Fleischer, D.: *Digital Circuit Elements; Switching Elements, Storage Elements*, 3rd ed., Siemens AG, Berlin, Munich (1979).

Fröhr, F.: *Electronic Controllers for Drive and Power Engineering*, 2nd ed., Siemens AG, Berlin, Munich (1977).

Fröhr, F., and Orttenburger, F.: *Introduction to Electronic Control Engineering*, 5th ed., Siemens AG, Berlin, Munich (1981).

Garten, W.: *Lead Storage Batteries*, Varta Series of Technical Books, Vol. 1, Varta Batterie AG, Hannover, VDI Verlag GmbH, Düsseldorf (1974).

Gerlach, W.: *Thyristors*, Springer-Verlag, Berlin, Heidelberg, New York (1981).

Hartel, W.: *Power Converter Circuits; Introduction to the Circuits of Mains-Operated Power Converters*, Springer-Verlag, Berlin, Heidelberg, New York (1977).

Herhahn, A.: *First Reader on Safety for Electric Installations; to VDE 0100 with Safety Glossary*, Vogel-Verlag, Würzburg (1978).

Heumann, K.: *Principles of Power Electronics*, Verlag B. G. Teubner, Stuttgart (1975).

Heumann, K., and Stumpe, A.: *Thyristors; Properties and Application*, Verlag B.G. Teubner, Stuttgart (1974).

Hille, W., and Schneider, O.: *Technical Book on Electrical Professions*, Verlag B.G. Teubner, Stuttgart (1972).

Hirschmann, W.: *Electronic Circuits*, Siemens AG, Berlin, Munich (1982).

Hoffmann, A., and Stocker, K.: *Thyristor Manual*, 4th ed., Siemens AG, Munich (1976).

Integrated Phase-Angle Control Systems TCA 780 for Power Electronics: Technical Reports, Components Division, Siemens AG, Berlin, Munich (1978).

Integrated Switching Power Pack Control Circuits TDA 4700/TDA 4718: Technical Reports, Components Division, Siemens AG, Berlin, Munich (1980).

Kinzelbach, R.: *Alkaline Batteries*, Varta Series of Technical Books, Vol. 3, Varta Batterie AG, Hannover, VDI Verlag GmbH, Düsseldorf (1974).

Kloss, A.: *Power Electronics without Padding; A Practical Introduction to Circuit Engineering in Power Electronics*, Franzis-Verlag, Munich (1980).

Kolb, O.: *Introduction to Power Converter Engineering*, Vol. 1: *Principles*, AT-Fachverlag GmbH, Stuttgart (1976).

Krakowski, H.: *Telecommunications Power Supply*, Telecommunications Engineering Textbook, Verlag Schiele & Schön GmbH, Berlin (1978).

Kurscheidt, P.: *Power Electronics*, Berliner Union Verlag, Stuttgart, and Verlag W. Kohlhammer, Stuttgart, Berlin, Cologne, Mainz (1979).

Lappe, R., *et al.*: *Thyristor Power Converters for Drive Control Systems*, VEB-Verlag TECHNIK, Berlin (1975).

Möltgen, G.: *Mains-Operated Power Converters with Thyristors; Introduction to Mode of Operation and Theory*, 3rd ed., Siemens AG, Berlin, Munich (1974).

Müller-Schwarz, W.: *Principles of Electronics*, 3rd ed., Siemens AG, Berlin, Munich (1974).

MWM Diesel: *Generating Sets Manual; Guide for the Planning and Installation of Fixed Diesel-Electric Systems*, Motoren-Werke AG, Mannheim (1975).

Pabst, D.: *Operational Amplifiers; Principles and Application Examples*, Dr Alfred Hüthig Verlag, Heidelberg (1976).

Patzschke, U.: *Applied Thyristor Engineering; Principles and Application of Power Electronics*, Telekosmos-Verlag, Franckh'sche Verlagshandlung, Stuttgart (1970).

Paul, R.: *Semiconductor Diodes; Principles and Application*, Dr Alfred Hüthig Verlag, Heidelberg (1976).

Paul, R.: *Transistors and Thyristors; Principles and Application*, Dr Alfred Hüthig Verlag, Heidelberg (1977).

Schilling, W.: *Thyristor Engineering; An Introduction to the Application of Semiconductors in Power Engineering*, Verlag R. Oldenbourg, Munich, Vienna (1968).

Schlotheim, G.: *Power Electronics; Tasks and Solutions*, Vogel-Verlag, Würzburg (1980).

Tholl, H.: *Components in Semiconductor Electronics*; Part 2: *Field-Effect Transistors, Thyristors and Optoelectronics*, Verlag B.G. Teubner, Stuttgart (1978).

Watzinger, H.: *Application of Mains-Operated Power Converters for Direct Current Drives*, Siemens AG, Berlin, Munich (1976).

Watzinger, H.: *Mains-Operated Power Converters with Direct Current Output; Circuits and Mode of Operation*, Siemens AG, Berlin, Munich (1972).

Watzinger, H.: *Power Converters – Direct Current Drives; Measuring, Recording, Fault Finding in Start-up and in Operation*, Dr Alfred Hüthig Verlag, Heidelberg (1980).

Westphal, H.: *Selenium Today*, ITT Components, Verlag Gruppe Europa (1973).

Witte, E.: *Lead–Acid and Alkaline Accumulators*, Varta Series of Technical Books, Vol. 4, Varta Batterie AG, Hannover, VDI Verlag GmbH, Düsseldorf (1977).

Wolf, G.: *Digital Electronics; The Mode of Operation of Integrated Logic and Storage Elements*, Franzis-Verlag, Munich (1977).

Wüstehube, J., *et al.*: *Switching Power Packs; Principles, Design, Examples of Circuits*, expert-Verlag, Grafenau/Württ, and VDE-Verlag, Berlin (1979).

Zach, F.: *Power Electronics; Components, Power Circuits, Control Circuits, Influences*, Springer-Verlag, Vienna, New York (1979).

Journals

Baer, G., and Wollschläger, M.: Power supply equipment for teleprint and data switching, *telcom report*, **3**, 203–206 (1980).

Becker, H., and Schmalzl, F.: Small power supply systems independent of the mains, *telcom report*, **3**, 212–217 (1980).

Engert, S., and Weller, K.: Construction and fault finding of power supply equipment for long-distance communication systems, *telcom report*, **3**, 48–51 (1980).

Forstbauer, W., Kublick, C., Schultze, W., Schwarz, R., and Vau, G.: Alternative power supply for all applications, Extended Reprint from *Siemens Journal*, **52**, July 1978, No. 7, 393–444 (1978).

Ganzer, E., Knesewitsch, J., and Ziegler, A.: Power supply equipment for long-distance communication systems, *telcom report*, **3**, 207–212 (1980).

Herfurth, M.: Selection circuits for SIPMOS transistors in switching mode, *Siemens components*, **18**, No. 5, 218–224 (1980).

Hermanspann, F., and Probst, H.: Power supply equipment for public and private telephone switching technology, *telcom report*, **3** 197–202 (1980).

Hertneck, K., Kübler, K., and Ponzer, F.: Central power supply for telecommunications systems, *telcom report*, **3**, 190–196 (1980).

Krakowski, H.: Telecommunications power supply, *Fernmelde-Ing.*, **1972**, Nos 9 and 10 (1972).

Krakowski, H.: Uninterrupted power supply for EDP systems, *etz-b*, **27**, No. 2, 39–41 (1975).

Krakowski, H., and Schott, H.: Power supply – an important factor for the reliable operation of world-wide communications links, *telcom report*, **3**, 185–189 (1980).

Neugebauer, K., and Schwarz, R.: Uninterrupted power supply for EDP systems; increasing the reliability and availability of EDP systems by means of an uninterrupted power supply, *etz*, **100**, No. 12, 606–610 (1979).

Rambold, K.: Power supply equipment for communications engineering – special features and applied principles, *telcom report*, **1**, 404–409 (1978).

Specifications and Standards

VDE 0100 Conditions for the equipping of power plants with rated voltage up to 1000 V.

VDE 0160 Part 1: Conditions for fitting out power plants with electronic equipment; Part 1: Installations with electronic equipment for information processing in power plants.

VDE 0160 Part 1b: Conditions for fitting out power plants with electronic equipment; Part 1: Installations with electronic equipment for information processing in power plants.

VDE 0556 Conditions for polycrystalline rectifiers.

VDE 0557 Conditions for single-crystal rectifiers.

VDE 0800 Part 1: Conditions for equipping and operating telecommunications systems, including information processing systems; Part 1: General conditions.

VDE 0800 Part 1c: Conditions for equipping and operating telecommunications systems, including information processing systems; Part 1: General conditions.

VDE 0800 Part 3: Conditions for equipping and operating telecommunications systems, including information processing systems; Part 3: Special conditions for systems with remote power supply.

VDE 0804 Conditions for telecommunications equipment, including information processing equipment.

DIN 19226 Automatic control engineering; definitions and names.

DIN 40108 Electrical power engineering; current systems, definitions, magnitudes, symbols.

DIN 40110 Alternating quantities.

DIN 40700 Part 8: Graphical symbols; semiconductor components.

DIN 40700 Part 14: Graphical symbols; digital information processing.

DIN 40700 Part 22: Graphical symbols; digital information processing, memory logic elements.

DIN 40706 Graphical symbols; power converters.

DIN 40729 Voltaic secondary cells (accumulators); definitions.

DIN 40736 Part 1: Lead storage batteries; fixed cells with positive box-type plates, cells in plastic containers, capacities, main dimensions, weights.

DIN 40736 Part 2: Lead storage batteries; fixed cells with positive box-type plates, cells in hard rubber containers, capacities, main dimensions, weights.

DIN 40737 Part 1: Lead storage batteries; fixed batteries with positive box-type plates, batteries in hard rubber monobloc containers, capacities, main dimensions, weights.

DIN 40737 Part 2: Lead storage batteries; fixed batteries with positive box-type plates, batteries in plastic monobloc containers, capacities, main dimensions, weights.

DIN 40738 Lead storage batteries; fixed batteries with positive large-area plates, close-fitting, capacities, main dimensions, weights.

DIN 41740 Part 1: Selenium diodes; definitions.

DIN 41740 Part 2: Selenium diodes; marking and general recommendations for particulars in data sheets.

DIN 41740 Part 3: Selenium diodes; measuring and testing methods.

DIN 41745 Stabilized power supply equipment; definitions.

DIN 41750 Part 1: Power converters; definitions for semiconductor converters, construction.

DIN 41750 Part 2: Power converters; definitions for converters, types and designations.

DIN 41750 Part 3: Power converters; definitions for converters, commutating, drive, electrical quantities.

DIN 41750 Part 4: Power converters; definitions for converters, mains operated converters for rectifying and inverting.

DIN 41750 Part 4, Supplement: Power converters; definitions for converters, calculation instructions for mains-operated converters for rectifying and inverting.

DIN 41750 Part 5: Power converters; definitions for converters, self-operated converters.

DIN 41750 Part 6: Power converters; definitions for converters, load-operated converters.

DIN 41750 Part 7: Power converters; definitions for converters, trigger sets.

DIN 41751 Power converters; semiconductor converter stacks and converter equipment, types of cooling.

DIN 41752 Power converters; semiconductor converter equipment, performance characteristics.

DIN 41755 Part 1: Superpositions on a direct voltage, periodic superpositions, definitions, measurement method.

DIN 41756 Part 1: Power converters; loading of converters, modes, load groups.

DIN 41756 Part 2: Power converters; loading of converters, types of load with direct current.

DIN 41756 Part 3: Power converters; loading of converters, types of load with alternating current.

DIN 41760 Power converters; polycrystalline rectifier plates and stacks; definitions and rated values.

DIN 41761 Power converters; converter circuits, designations and identification characters.

DIN 41761 Supplement: Power converters; converter circuits, designations and identification characters; examples.

DIN 41762 Part 1: Power converters; power characteristics for semiconductor converter stacks; polycrystalline rectifier stacks.

DIN 41762 Part 2: Power converters; power characteristics for semiconductor converter stacks; single-crystal converter stacks.

DIN 41772 Power converters; semiconductor rectifiers, shapes and abbreviations of characteristics.

DIN 41772 Supplement 1: Power converters; semiconductor rectifiers, examples of characteristic for chargers.

DIN 41772 Supplement 2: Power converters; semiconductor rectifiers, examples of characteristic for equipment in parallel mode with batteries.

DIN 41774 Power converters; semiconductor rectifiers with W characteristic for the charging of lead storage batteries, recommendations.

DIN 41777 Part 1: Power converters; semiconductor rectifiers and systems, basic requirements for the maintenance charging of lead storage batteries.

DIN 41777 Part 2: Power converters; semiconductor rectifiers and systems, controlled and regulated equipment with constant voltage characteristic for the maintenance charging of lead storage batteries.

DIN 41777 Part 3: Power converters; semiconductor rectifiers and systems, unregulated equipment with W characteristic for the maintenance charging of lead storage batteries.

DIN 41781 Rectifier diodes for power electronics; definitions.

DIN 41782 Rectifier diodes; recommendations for data sheet particulars.

DIN 41783 Single-crystal rectifier cells; measuring methods.

DIN 41784 Part 1: Thyristors; measuring and test methods.

DIN 41785 Part 1: Semiconductor components; abbreviations for use in data sheets, construction of the abbreviations.

DIN 41785 Part 2: Semiconductor components; abbreviations for use in data sheets, abbreviations for semiconductor components in communications engineering.

DIN 41785 Part 3: Semiconductor components; abbreviations for use in data sheets, abbreviations for semiconductor components in power electronics.

DIN 41785 Part 4: Semiconductor components; abbreviations for use in data sheets, abbreviations for digital binary microcircuits.

DIN 41785 Part 5: Semiconductor components; abbreviations for use in data sheets, abbreviations for linear integrated amplifiers.

DIN 41786 Thyristors; definitions.

DIN 41787 Thyristors; recommendations for data sheet particulars.

DIN 41789 Marking of rectifier diodes and thyristors.

DIN 41791 Part 1: Semiconductor components for communications engineering; particulars in data sheets, general.

DIN 41791 Part 5: Semiconductor components for communications engineering; particulars in data sheets, power transistors.

DIN 41791 Part 6: Semiconductor components for communications engineering; particulars in data sheets, switching transistors.

DIN 41794 Part 9: Reliability details for semiconductor components; thyristors.

DIN 41852 Semiconductor technology; definitions.

DIN 41855 Semiconductor components and integrated circuits; types and general definitions.

DIN 57105 Part 1/VDE 0105 Part 1: VDE regulation for the operation of power plants; general conditions.

DIN 57160 Part 2/VDE 0160 Part 2: VDE regulation for the equipping of power plants with electronic equipment; installations with power electronics equipment in power plants.

DIN 57411 Part 1/VDE 0411 Part 1: VDE regulation for electronic measuring instruments and controllers; Protective measures for electronic measuring equipment.

DIN 57411 Part 1a/VDE 0411 Part 1a: VDE regulation for electronic measuring equipment and controllers, part amendment a [VDE Regulation].

DIN 57510/VDE 0510: VDE regulation for lead storage batteries and battery installations.

DIN 57558 Part 1/VDE 0558 Part 1: VDE regulation for semiconductor power converters; general conditions and special conditions for mains-operated converters.

DIN 57558 Part 1a/VDE 0558 Part 1a: VDE regulation for semiconductor power converters; general conditions and special conditions for mains-operated converters, part amendment a [VDE Regulation].

DIN 57558 Part 2/VDE 0558 Part 2: VDE regulation for semiconductor power converters; special conditions for self-operated converters.

DIN 57558 Part 3/VDE 0558 Part 3: VDE Regulation for semiconductor power converters; special conditions for d.c. chopper controllers.

DIN 57800 Part 1d/VDE 0800 Part 1d: Regulations for setting up and operating telecommunications systems, including information processing systems; Part 1: General conditions, part amendment d.

DIN 57800 Part 2/VDE 0800 Part 2: Telecommunications engineering; earthing and potential equalization [VDE Regulation].

DIN 57804/VDE 0804: Telecommunications engineering; manufacture and testing of equipment.

DIN 57871/VDE 0871: Radio noise suppression of high-frequency equipment for industrial, scientific, medical and similar purposes [VDE Regulation].

DIN 57874/VDE 0874: VDE guidelines for measures for radio noise suppression.

DIN 57875/VDE 0875: VDE regulation for the radio noise suppression of electrical equipment and systems.

Index

91 56